CELL POTASSIUM

TRANSPORT IN THE LIFE SCIENCES

EDITOR: E. EDWARD BITTAR

VOLUME 1   *Cell Potassium* — Roderick P. Kernan

CELL POTASSIUM

Roderick P. Kernan, D.Sc., Ph.D., M.R.I.A.
Department of Physiology
University College, Dublin

A Wiley-Interscience Publication
JOHN WILEY & SONS
New York · Chichester · Brisbane · Toronto

Copyright © 1980 by John Wiley & Sons, Inc.

All rights reserved. Published simultaneously in Canada.

Reproduction or translation of any part of this work beyond that permitted by Sections 107 or 108 of the 1976 United States Copyright Act without the permission of the copyright owner is unlawful. Requests for permission or further information should be addressed to the Permissions Department, John Wiley & Sons, Inc.

Library of Congress Cataloging in Publication Data:

Kernan, Roderick P
   Cell potassium.

  (Transport in the life sciences ; v. 1)
  "A Wiley/Interscience publication."
  Includes index.
  1. Cell metabolism. 2. Potassium metabolism.
I. Title. II. Series.

QH634.5.K47   574.87′61   80-13332
ISBN 0-471-04806-2

Printed in the United States of America

10 9 8 7 6 5 4 3 2 1

To Mary, Sue and Niall

SERIES PREFACE

Membrane transport is rapidly becoming one of the best-worked fields of modern cell biology. The Transport in the Life Sciences series deals with this broad subject in monograph form. Each monograph seeks to trace the origin and development of ideas in the subject in such a way as to show its true relation to membrane function. It also seeks to present an up-to-date and readable outline of the main problems in the subject and to guide thought on to new lines of investigation.

The task of writing a monograph is not a light one. My special gratitude is to the various authors for expounding their subjects with scholarly care and force. For the preparation of the indexes I thank Dr. Barbara Littlewood.

E. Edward Bittar

Madison, Wisconsin
June 1980

PREFACE

Since the publication of *Cell K* by Butterworth Inc. in 1965, the literature on the role of this cation in biological processes has grown considerably. The necessity to deal with the relationship between potassium and other cations, including sodium, calcium, and hydrogen in regard to membrane fluxes and permeability changes, not to mention cell metabolism in general, has tended to broaden the scope of the book but restricted the depth of treatment of the subject. Some selectivity was also required taking into consideration current trends based largely on technical developments. Three topics dealt with in greater detail than in the previous volume are epithelial transport, acid–base balance, and mitochondria in relation to potassium equilibria and movements across biological membranes. The recent award of the Nobel Prize in Chemistry to Peter Mitchell of Cornwall for his original and provocative chemiosmotic hypothesis of energy conservation in mitochondria has perhaps contributed to the topicality of the chapter devoted to electrolyte metabolism in mitochondria. In relation to epithelial potassium transport, a question of some clinical significance is the mechanism of action of some diuretics, including the potassium-conserving amiloride, on the kidney tubule, and this is still being investigated largely through the use of potassium-selective microelectrodes which can monitor the intracellular potassium of the epithelial cells.

For permission to reproduce illustrations and electronmicrographs, my thanks are due to the many authors concerned and to the Editors of the following: *The Journal of Physiology, Journal of General Physiology, Journal of Cell Biology, Journal of Experimental Biology, Journal of Membrane Biology, Nature* (London), *Pflügers Archives, Biochimica Biophysica Acta, Archives of Biochemistry and Biophysics, European Journal of Biochemistry, Annals of the New York Academy of Sciences, Journal of Experimental Botany, Journal of Clinical Pathology, Proceedings of the Royal Irish Academy,* the *Physiological Reviews,* and the *Quarterly Review of Biophysics.* I am also indebted to Butterworths for permission to undertake the present work. Finally I wish to thank Mr. P. Auerbach for help with photography and Mrs. Jill Brennan for typing the manuscript.

RODERICK P. KERNAN

Dublin, Ireland
April 1980

CONTENTS

1 Assimilation from the Environment 1
   Potassium Uptake in Algae, 5
   Potassium Uptake and Metabolism of Fungi, 10
   Potassium Uptake in Higher Plants, 11

2 The Measurement of Cellular Potassium 17
   Analysis of Tissue Biopsies, 18
   Whole Body Counters, 20
   Exchangeable Potassium Measurement, 23
   Direct Measurement of Intracellular Potassium by Means of Ion-Selective Microelectrodes, 25

3 Potassium Equilibrium and the Resting Cell 31
   Association–Induction Hypothesis, 32
   The Membrane Theory, 38
   Anomalous Rectification in Muscle, 45
   Blocking of Potassium Channels, 51
   Factors Increasing Potassium Conductance, 54

4 Potassium Fluxes and the Action Potential 58
   Hodgkin–Huxley Model of Conductance Changes During the Action Potential in Nerve, 59
   The Action Potential in Skeletal Muscle, 62
   Gating Current for Potassium Channels, 66
   Potassium Current in the Repolarization of Cardiac Muscle, 70

Gating and the Effects of Calcium and Hydrogen Ions, 80

Conclusions, 82

5 Active Transport of Potassium in Nonepithelial Cells — 85

General Structure of Transport Protein, 86

The Sodium–Potassium Carrier of Red Cell "Ghosts", 89

Potassium–Potassium Exchange and Dephosphorylation, 92

Low- and High-Potassium Red Blood Cells, 93

Conformational Changes in Transport Enzyme Produced by Sodium and Potassium Ions, 96

Potassium–Activated Phosphatase and the Sodium–Potassium Pump, 99

Active Transport in Skeletal Muscles, 100

Electrogenic Sodium Pump and Potassium Uptake, 102

Effect of Innervation on Potassium Transport and Permeability, 103

Stimulation of the Sodium Pump by Insulin and by β-Adrenergic Agents, 104

Potassium and Sodium Transport in Cardiac Tissue, 106

Potassium Transport in Smooth Muscle, 107

Sodium and Potassium Transport in Dialyzed Squid Axons, 109

Potassium Transport in Liver, 112

Potassium Uptake in Yeast and Bacteria, 112

6 Potassium Fluxes in Mitochondria — 120

The Membrane Potential in Mitochondria, 121

Potassium Ions and the Chemiosmotic Hypothesis, 124

Volume Control and Potassium Transport in Mitochondria, 131

Potassium Fluxes and Uncoupling Agents, 132

7 Potassium Ions in Cell Metabolism   **138**

Potassium in Protein Synthesis, 140

Potassium Deprivation and Enzyme Changes in Mammals, 143

The Role of Potassium in Purine Biosynthesis, 145

Other Reactions Catalyzed by Potassium and Their Characterization, 146

8 Active Transport of Potassium in Epithelial Tissues   **151**

Koefoed–Johnsen Ussing Model of Transepithelial Transport, 152

Measurement of Intracellular Potassium Activity and of Relative Ion Permeabilities, 154

Electrogenic or Neutral Pump, 154

Transepithelial Potassium Transport, 159

Potassium Transport in the Cochlea, 160

Potassium Transport in Salivary Glands, 165

Potassium Transport in Sweat Glands, 167

Renal Potassium Transport, 167

Diuretics and Renal Potassium Movement, 171

Conclusions, 172

9 Cellular Potassium and Acid–Base Balance   **175**

Intracellular pH of Muscles at Rest, 176

Active Transport and Control of pH_i, 181

Cellular Potassium in Acidosis and Alkalosis, 182

Index   **187**

CELL POTASSIUM

1
Assimilation from the Environment

The name potash is derived from an early source of the metal, namely, from the burning and leaching of wood and other vegetable matter and evaporation of the extract in pots yielding potassium carbonate. In order of abundance it is the eighth element, making up about 2.5% by weight of the lithosphere. Present originally within igneous rocks, including orthoclase feldspar ($6SiO_2 \cdot Al_2O_3 \cdot K_2O$, with 7–9% K) and the micas biotite [$K(MgFe)_3AlSi_3O_{10}(OH)_2$, 4.5–7% K] and white muscovite [$KAl_3Si_3O_{10}(OH)_2$, 6–9% K], potassium was released in soluble form in the processes of mechanical and chemical weathering, leaving behind insoluble clay minerals such as kaolinite ($Al_2O_3 \cdot 2SiO_2 \cdot 2H_2O$). Weathered biotite shows a gain of Al, Si, and H_2O, oxidation of Fe, and a loss of Mg^{2+} and K^+. Although disordered micas are most abundant in the clay fraction of soils, they occur also in silts, sands, and gravels. The potassium-containing minerals in which the element existed as its oxide were generally composed of sesquioxides of iron and aluminum coupled with silica. The relative amounts of the heavier metal oxides to silica usually determined the ease with which the potassium could be removed from the crystalline complex of the mineral.

Some micas were very resistant to chemical weathering while others released potassium when leached with carbonic acids in rain water. Other anions such as the halogens of volcanic origin[1] contributed to the chemical weathering of igneous rocks, changing them to metamorphic and sedimentary rocks and to clays, releasing the minerals needed for the generation of life on the planet. During and after the mesozoic era (165–160 Myr ago), the biosphere interacted to an increasing extent with the lithosphere, making it in some respects a more favorable environment for the growth and proliferation of plant life. For example, plants themselves contributed to the weathering process, in that hydrogen ions released from their root hairs exchanged for metallic ions, including potassium contained in the crystal lattices of minerals in the soil. The fibrous root systems of decaying plants combined with the products of the chemical weathering of igneous rocks to form clay–humus complexes, which in colloidal form behaved as soil buffers and sources of potassium ions.

In the oceans the ratio of concentration of sodium to potassium was about 30 to 1, resembling that found in blood plasma and interstitial fluid of the mammals and amphibians. While the concentrations of these cations were similar in igneous rocks, the concentration ratio K/Na in runoff water seems to be proportional to the volume of water available for dissolution of the rock. These ions were probably carried to the oceans in equal quantities, but potassium to be found there decreased with time. The explanation of this is probably to be found in the composition of pelagic sedimentary rocks, which contain about three times more potassium than sodium. Now while much of the missing potassium was probably bound by cation-deficient clays washed into the seas in fine suspension before life appeared in the ocean, the contribution of marine microorganisms and of plant and animals life to potassium fixation must have been considerable. Potassium absorbed preferentially from seawater during the hundred of millions of years since life first appeared in the oceans has fallen to the sea floor trapped in the organic debris, there undergoing silicification and calcification to metamorphic rocks such as glauconite.[2] In this way the element has run through a cycle from mineral to mineral by way of the living cell.

This may have been a major factor in the evolution of the oceans during the Mesozoic era.[3] The sedimentary deposits may reach a depth of 100 m in places, as found by the D. V. Glomar Challenger[4] in it drilling of cores from the ocean bed. The contribution of organic matter to the upper levels of this sediment may be judged by the contents of a dredge of the Central Pacific made by H.M.S. Challenger (1872–1876) which included 1500 shark teeth and about 50 ear bones of whales.[5] When the amphibia moved to dry land, they took their watery environment with them as an "internal sea" of plasma and extracellular fluid in which the concentration ratio of Na to K was similar to that found in the oceans. We may wonder about the electrolyte composition of the "primeval broth"[6] in which life first appeared on our planet and to what extent it determined the permeability properties of the cell membrane. If the oceans had been rich in potassium, would life have evolved at all or would cells have remained virtually impermeable to potassium?

Of course, primitive cells including bacteria and yeasts which had to contend with wide variations in external salt concentrations and external osmotic pressures generally have very low potassium permeability, and they have also evolved a tough cell wall as a protection against the disruptive effect of increasing turgor pressure. During log growth phases in such cells when potassium is being rapidly accumulated, this turgor pressure is much higher than in the stationary phase, and it is believed that the pressure is necessary for normal growth and division.[7] Unlike the higher animals, these unicellular organisms show wide variations in the potassium concentration of their cells. This may be almost identical with that of the bathing fluid in resting cells, increasing above 500 mM[8] in the presence of nutrient and at a favorable external pH value. Plant cells also can withstand wide variation of turgor pressure because of their cellulose walls but their uptake of potassium is determined by the processes of photosynthesis, transpiration, and the salt conservation and metabolism of the plant as a whole. In this respect the metabolism of the root hair and the base exchange character of the soil particles surrounding it have an important role. In the micaceous earths and feldspars, potassium is held in hexagonal cavities formed by oxygen atoms (Fig. 1.1) between the lattices of the crystalline structure.[9] The cation binding capacity resides in the excess negative charge on the lattice. The penetration of H_2O and H_3O^+ into some of the interlayer cavities of the micacious earths through "frayed edges" leads to replacement of potassium in the lattice by hydrogen ions excreted by roots and leads to a loosening of its structure, exposing more potassium to the action of the environment. The mineral becomes potassium depleted changing in the process through the stages mica $\rightleftharpoons$ illite $\rightleftharpoons$ vermiculite $\rightleftharpoons$ montmorillonite. Reactions to the right lead to lattice exposure and lowering of K^+ activity. Drying (or freezing) of the mica of soil hastens potassium release if the exchangeable potassium level is low. Addition of high levels of soluble K^+ to soil by mass action tends to drive the reaction to the left, a change which is also facilitated by drying or freezing. Ammonium ions tend to prevent cleavage as they occupy cavities normally used by K^+. The concentration of potassium in soil water generally varies from 6 to 20 mmole·l^{-1}, being maintained within these limits by release of cation from absorption on clay particles on the one hand and by absorption into root hairs in the rhizosphere on the other. With the exception of the black mica biotite,

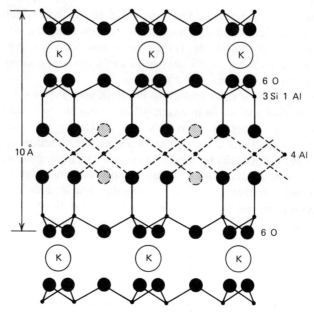

Figure 1.1 Schematic presentation of the molecular sheet structure of the disilicate muscovite, showing the location of potassium in the crystal lattice. From R. P. Kernan, *Cell K*, Butterworths, 1965.

found in granite, the contribution of primary igneous rocks to potassium replenishment in the biosphere is minimal. The clays either with or without associated humus provide the main reservoir of potassium. Their extraordinarily high exchange capacity may be judged from their large surface area which is available to cation adsorption. Vermiculite, for example,[10] has an external surface area of 39 m$^2 \cdot$ g^{-1} and an internal lattice area of 511 m$^2 \cdot$ g^{-1}, with a total exchange capacity of about 140 mmole $\cdot$ 100 g^{-1} mineral. It has been calculated that 1 mm^2 of root hair surface makes contact with 10^8 clay particles each carrying 6000–7000 exchangeable cations, so the possibility of direct exchange of H$^+$ for K$^+$ between root hair and soil colloid without the significant participation of soil water has been considered.[11]

Under conditions of intensive agriculture, especially with high-yield varieties, even this potassium reserve would soon be depleted without the addition of potassium fertilizers to the soil. A large fraction of such potassium becomes fixed within the clay particles in a nonexchangeable form, thereby preventing its leaching by rain water. The expanding–contracting property of micaceous earths like montmorillonite and the geometric arrangement of oxygen ions at the surface of the crystal layer are the basis of their ion exchange properties. The oxygens are arranged hexagonally, the opening within the hexagon being 2.8 Å in diameter. As the clay is dehydrated, the layer contracts and the ions lose their hulk of oriented water molecules approaching the unhydrated ionic diameter size.

Ions whose diameters fit snugly into the lattice "holes" should be held tightly because they are closer to the anionic charges within the crystal; by fitting into the "holes" they allow the layers to come closer together and be locked against rehydration and reexpansion. Larger ions which cannot enter the "holes" remain loosely held between the layers rather than within the layer thereby facilitating rehydration. Small ions, on the other hand, while entering the holes are too small to bind the two layers together (and the dielectric constant may be about unity).[12] The maximum spacing was 12 Å for potassium-saturated materials, whereas Ca^{2+}-, Mg^{2+}-, Na^+-, and H^+-saturated samples had a spacing of about 16 Å, equivalent to a double layer of water molecules. Fixation of K^+ by montmorillonite does not occur in the absence of drying. Physical conditions, such as alternating wetting and drying of soil coupled with chemical changes such as the liberation of organic matter into the rhizosphere, tend to increase the amount of the fixed potassium which may be recoverable or available to the plant roots.

POTASSIUM UPTAKE INTO ALGAE

Large coenocytic cells which may reach a diameter of 1 cm have been used in the study of potassium uptake in plants. These cells have large vacuoles surrounded by streaming protoplasm about 10 μm thick. Uptake of potassium by such cells may be considered as a transfer across two membranes in series, the first being the plasmalemma separating external fluid and cytoplasm and the second being the tonoplast between cytoplasm and vacuole. So in order to establish the site of active transport of ions, it is necessary first to determine the ionic status of both intracellular fluids. Cells have been centrifuged[13] to produce a clear demarkation between cytoplasm and vacuolar fluid and then pinched along this line, and the two fractions were separated for chemical analysis. It has been found[14] that in freshwater algae cytoplasm usually contains more potassium than the vacuole, while the reverse is true of marine algae. The membrane potentials at plasmalemma and tonoplast was then measured so that the electrochemical potential of potassium, sodium, and chloride in each fluid compartment could be calculated. In the case of *Nitella translucens*,[15] a freshwater alga (Fig. 1.2), it is evident that while the cytoplasm may contain about 1110 times more potassium than the exter-

6 Assimilation from the Environment

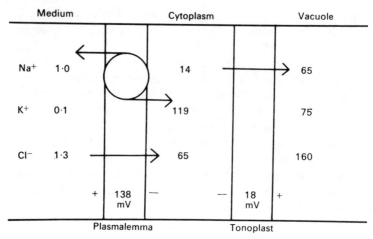

Figure 1.2 Ionic and electrical gradients across membranes of *Nitella translucens*. Active pumping indicated by arrows. Concentrations in m.mole·kg⁻¹ water. From R. M. Spanswick & E. J. Williams, *J. Exp. Bot.* **15**, 193 (1964).

nal fluid which resembles pond water, the internal negativity of 138 mV might go a long way toward preventing the loss of this cation from the cytoplasm to the external fluid. The membrane potential at which potassium ions would be at the same electrochemical potential on both sides of the plasmalemma is defined by the Nernst equation,

$$E_K^{oc} = \frac{RT}{zF} \ln \frac{a_K^o}{a_K^c} \tag{1}$$

where E_K^{oc} is the potential within the cytoplasm relative to external fluid, and a_K^o and a_K^c are the activities of potassium ions in external fluid and cytoplasm, respectively. On the assumption that the activity coefficient for potassium ions is the same in both fluids, it is usual to substitute concentration for activity, which on the basis of Fig. 1.2 and a temperature of 25° would give the following result:

$$58 \log \frac{0.1}{119} = -178 \text{ mV} \tag{2}$$

Since the measured electrical potential across the plasmalemma appeared to be less negative than E_K^{oc} by about 40 mV, potassium ions may be expected to leave the cell if the plasmalemma is permeable to these ions. Similar measurements have been made in the case of the other ions and for both membrane, yielding differences between measured potentials and calculated equilibrium potentials for Na⁺, K⁺, and Cl⁻ of +56, −5, and +4

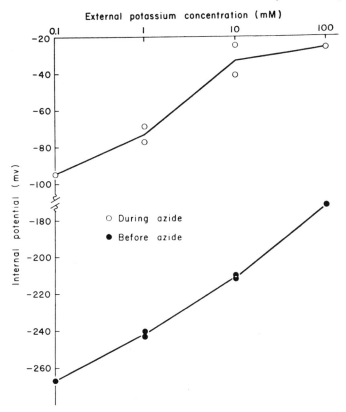

Figure 1.3 Effect of external potassium on E_m of Neurospora in presence and absence of 1 mM sodium azide. From C. L. Slayman, *J. Gen. Physiol.* **49,** 93 (1965).

mV, respectively, at the tonoplast and of −72, +40, and −236 mV, respectively, at the plasmalemma. Such results lead to the conclusion that chloride is actively transported into the cytoplasm and probably moves passively from there to the vacuole. Sodium appears to be pumped from the cytoplasm both into the vacuole and to the exterior. Potassium appears to be actively pumped into the cytoplasm from the external fluid but to be at electrochemical equilibrium across the tonoplast allowing for experimental error. Activity measurements with ion-selective electrodes[14] seem to verify the use of concentrations in place of activities, but more recent experiments with *Nitella* have raised doubts regarding the potassium permeability of the plasmalemma under the conditions described. Membrane permeability to potassium seems to depend on a number of factors, including the pH of the external fluid,[16] the presence or absence of calcium ions[17] in this fluid, and the external potassium concentration. There seems little doubt that potassium permeability is increased on increasing $[K]_0$ and depolarization of the plasmalemma may be produced thereby. Removal of calcium[18] from the medium also brought about better agreement between E_K^{oc} and the plasmalemma potential, suggesting passive distribution and movement of potassium across the membrane. Furthermore, it has been found[16] that at 0.5 mM potassium and pH 6, electrochemical equilibrium also exists for potassium ions across the cell membrane. Above and below this pH value, however, membrane depolarization seemed to take place, with a consequent tendency for potassium to be lost from the cells. The experiments just described were carried out in the dark because light produced membrane hyperpolarization which was probably due to the electrogenic pumping of ions across the two membranes. For example, the active transport of chloride from external fluid to cytoplasm or of sodium or hydrogen ions in the opposite direction could be expected to make the inside of the cells more negative.

Although there have been indications that potassium ions may be accumulated passively by freshwater algae including *Chara* and *Nitella*, in the case of the latter there remain some experimental findings which are difficult to reconcile with this view. There was the absence of cell depolarization on increasing $[K]_0$ in the presence of calcium already referred to. On the other hand, under certain conditions the cell membrane behaves like an H$^+$ electrode,[19] suggesting high permeability to these ions. While it is true that calcium and hydrogen ions in certain tissues, including muscle,[20] may intereact with permeability channels in the membrane thereby modifying their selective permeability towards cation with respect to anions, in *Nitella* the P_K and P_{Cl} in the membrane appeared to be independent of pH. It is therefore unlikely that the membrane depolarization produced by hydrogen ions is mediated through permeability change to other ions.

Membrane fluxes and electrical conductance across the membranes of *Nitella translucens* and *Chara corallina* have also been measured[21] with a view to determining relative ion permeabilities and to compare total membrane conductance with "flux conductance" determined from measurement of the unidirection fluxes of ions following a change in membrane potential imposed by a voltage clamp. The relationship for monovalent ions at equilibrium between these is given by the equation

$$g_j = \left(\frac{F^2}{RT}\right) m_j \qquad (3)$$

where g_j is the conductance by ions j and m_j is the unidirectional flux of these ions across the membrane. The partial ion conductances for K$^+$,

Na$^+$, and Cl$^-$ were found to be 7, 1, and 1 μmho·cm^{-2} in *Nitella*, compared with total membrane conductance of 21 μmho·cm^{-2}. In *Chara*, partial ion conductances were 15, 1, and 2, respectively, compared with a total conductance of 70 μmho·cm^{-2}. There appeared to be a flux conductance deficit which might have been due to contribution by hydrogen ions. However eq. (3) is based on the assumption of independent diffusion through permeability channels, an assumption which has not been borne out in other tissues, including nerve;[22] it was therefore concluded that this principle did not hold but that there was interaction between fluxes during permeation through long channels. Another complication in such measurements is the presence of the cell wall outside the plasmalemma. This has a net negative fixed charge which gives rise to an uneven distribution of potassium and other ions between external fluid and wall. For example, potassium ion concentration in wall water is probably about 4.7 mM in the presence of $[K]_0$ of 1.1 mM.

In *Griffithsia*, a marine algae genus, measurements[14] of ion activity by means of ion-selective electrodes indicated an activity coefficient of 0.68 for cytoplasm, which was about the value which would be expected in seawater. Calculations based on the activity measurements led to the conclusion that potassium ions were at or near to electrochemical equilibrium across plasmalemma but that a significant difference of electrochemical potential gradient for potassium was present across the tonoplast, indicating that the cation would require to be actively transported from cytoplasm to vacuole. In the freshwater *Characeae*,[23] however, potassium activity measurement indicated its passive distribution across both membranes. The plasmalemma membrane of *Griffithsia* shows a very high selectivity for potassium over sodium, with a calculated P_{Na}/P_K ratio of 0.002–0.006, and this was coupled to a rapid response of PD across this membrane to changes in $[K]_0$. In artificial seawater (ASW) which contains 10 mM K and 490 mM Na, the electrical resistance of the plasmalemma was 150–260 Ω cm^{-2}, compared with 4000–5500 Ω·cm^{-2} at the tonoplast, so the main barrier to diffusion seemed to be at the latter. On increasing the external potassium concentration, the resistance of plasmalemma decreased while that of the tonoplast increased.

Illuminated algae generally had a membrane potential (E_m) which was about 50 mV greater than in the dark. Therefore under these conditions, E_m was more negative than E_K and, provided that the plasmalemma was permeable to potassium, it should have moved into the cell because of this potential difference. When membrane potential was clamped at E_K in the light, the current required was equivalent to about 20 pmole·cm^{-2}·sec^{-1}, which may be the current generated across the plasmalemma by the pumping of H$^+$ from the cell.

In *Halicystis ovalis*,[24] which has a particularly large vacuole making possible short-circuit current measurements and the use of multiple internal electrodes, it has been found that measured short-circuit current could be accounted for in terms of sodium transport to the external fluid (39.2 ± 5.4%) and of chloride transport into the vacuole 57.6 ± 5.3%) within the limits of experimental error, indicating that potassium did not need to be transported into the cell. *Halicystis parvula*, on the other hand, may be unique among algae in having similar concentrations of K^+, Na^+, and Cl^- in vacuolar sap[25] to those in seawater, yet with an internal negativity that can be accounted for by an inwardly directed active chloride transport.[26]

POTASSIUM UPTAKE AND METABOLISM OF FUNGI

In the mature hyphae of *Neurospora crassa* the internal potential often exceeded −200 mV and was sensitive to changes in external potassium and sodium concentrations.[27] For example, over the range of concentrations of 0.1 to 10 mM, potassium in the absence of other cation depolarized by 45 mV for a tenfold increase in concentration. A slope of 30 mV/log unit was obtained in the case of sodium. As in the case of bacteria,[28] the intracellular cation concentrations varied greatly, depending on whether the hyphae were resting or in the log phase of growth. In the former, $[K]_i$ was 146 ± 3 m.mole·kg^{-1} cell fluid and $[Na]_i$ was 26 ± 3 m.mole·kg^{-1}, while in the latter the values were 180 ± 3 and 14.0 ± 0.5 m.mole·kg^{-1} cell fluid, respectively. Also, like bacteria and yeasts,[29] potassium was lost only very slowly into distilled water, that is, at the rate of about 1.2 m.mole·hr^{-1}, although this rate of loss was increased 20-fold when sodium was added to the external fluid. This observation, coupled with the marked depolarizing action of sodium on the hyphae, indicated a relatively high permeability to this cation and the need for an effective active sodium excretion to maintain this cation at low concentrations within the cell. In contrast to the algae in which chloride was pumped inward and seemed to be the principal counterion to potassium, in *Neurospora* $[Cl]_i$ was only about 0.2 m.mole·kg^{-1} cell fluid, so internal phosphate was probably more significant.

The membrane potential in *Neurospora*[30] appeared to be made up of two parts (Fig. 1.3), one dependent directly on metabolism, the other a diffusion potential. In 10 mM potassium, calculated E_K was −73 mV, while measured potential was −140 mV and the addition of cyanide produced a 75% depolarization with an exponential time constant of 5.7 sec closely following the fall in intracellular ATP concentration. More recent experiments with this fungus and with its respiratory-deficient mutant[31] has led to the conclusion that a significant fraction of the membrane potential may be generated by the active transport of hydrogen ions out of the hyphae, the resulting high transmembrane potential providing the energy for the passive uptake of potassium ions into the cell. A leak of the hydrogen ions back into the hyphae is also believed to be responsible for a cotransport of nonelectrolytes, including glucose, as has also been found in yeast,[32,33] which also actively excretes hydrogen ions in large quantities.

The mutant of *Neurospora crassa,* which has mitochondria defective in cytochromes b and aa_3, showed impaired ability to accumulate potassium.[34] When grown in normal medium it contained 143 ± 15 m.mole·kg^{-1} cell water of potassium, compared with 180 ± 16 m.mole·kg^{-1} in the wild strain; and the mutant required 0.9 mM $[K]_0$ for adequate growth, compared with 0.3 mM or less in normal strain.

POTASSIUM UPTAKE IN HIGHER PLANTS

What has been found in relation to potassium accumulation in algae and fungi seems to be relevant to root systems and storage tissues of higher plants also. Naturally because of transpiration the latter are more complicated. For this reason transport here has been studied mainly in excised roots or detopped plants on the one hand and in isolated slices of storage tissue on the other. The higher plant has a vascular system and the sap within this has usually a potassium concentration intermediate between that of the soil water and parenchymal cells. The sap may contain[35] 11–18 mM potassium, compared with 100–200 mM potassium in the cell fluid. When microelectrodes were introduced into the root hairs of oat and pea seedlings, immersed in 0.1 to 3 mM potassium, the membrane potential was found to be in the range of -102 to -123 mV. Using the Nernst equation, the intracellular potassium concentrations that would be in equilibrium with the membrane potential were calculated for $[K]_0$ of 0.1, 1.0, and 3.0 mM potassium, respectively, and found to be 105, 355, and 1050 m.mole·l^{-1} cell fluid, respectively. In the case of the pea plant, the corresponding measured tissue potassium concentrations were 31.0, 75.4, and 69.0 m.mole·kg^{-1} fresh weight, indicating an inward driving force for potassium accumulation into the root hair cells. Since the potassium concentration in soil water is probably above this range and the bulk water flow associated with transpiration should also favor its uptake, it seems unlikely that much energy is required for its accumulation.

It has been stressed however that in the presence of a large inwardly directed flux of water and solute, simple Nernst criteria for passive ion movement may not be applicable and the use of the Ussing equation,[36,37] which requires the measurement of the flux ratio for potassium across the membrane, may be more valid, although independent migration of ions must still be assumed. This relationship may be stated as follows:

$$\frac{m_i}{m_0} = \frac{[K]_0}{[K]_i} \exp \frac{EF}{RT} \qquad (4)$$

where m_i and m_0 are potassium influx and efflux, respectively, and the activity coefficients for potassium are similar on either side of the cell membrane.

The above equation has been applied[38] to mineral contents and membrane potential of tissue of pea and oat seedlings and has indicated that of eight ionic species examined only potassium approached the relationship predicted for diffusion. Usually the potassium found within the cells was less than that predicted, probably because the membrane was hyperpolarized by electrogenic pumping of protons. More recent compartmental analysis[39] also using this equation appears to confirm the view that potassium may move passively into the roots.

However kinetic studies[40] on the rate of potassium uptake into excised barley roots in relation to external potassium concentration and that of other cations has indicated a carrier-mediated transport insofar as uptake rate appears to reach a plateau with increasing $[K]_0$ and hydrogen ions inhibit uptake in a competitive manner. The binding of transported cation (or anion) to carrier molecule in the cell membrane has been compared with the binding of substrate to enzyme; the principle of enzyme kinetics would therefore appear to be applicable to the accumulation of electrolytes and nonelectrolytes by cells. The assumption has been made that the rate (v) of transport of potassium into the cell was proportional to the amount of this cation bound to carrier molecule through the reaction

$$[K^+] + [M^-] = [K^+M^-] \qquad (5)$$

where $[M^-]$, the concentration of unoccupied carrier molecule within the membrane, is finite. As $[K]_0$ is increased, the quantity of potassium in the form $[K^+M^-]$ and consequently its rate of uptake should increase until $[M^-]$ becomes exhausted and the maximum uptake rate (V) is reached. The relationship between the reciprocal of $[K]_0$ and reciprocal of v, based on Lineweaver–Burk's treatment[41] of Michaelis–Menten kinetics as applied to enzyme kinetics, has been a convenient form in which to express competitive interaction of various ions on potassium uptake by root hairs, yeast cells, and other cells. This relationship may be expressed as

$$\frac{1}{v} = \frac{1}{[K]_0} \times \frac{k_{mK}}{V} + \frac{1}{V} \quad (6)$$

where k_{mK} is a constant similar to the Michaelis constant. When the reciprocal of the potassium uptake rate was plotted against the reciprocal of its concentration, a straight line was obtained with an intercept equal to the reciprocal of the maximum uptake rate V and with a slope of k_{mK}/V. Where another cation such as a proton competed with K^+ for binding with external carrier sites, the maximum uptake rate for potassium remained unchanged but the slope line changed in accordance with the following equation:

$$\frac{1}{v} = \frac{1}{[K]_0} \times \frac{k_{mK}}{V} \left(1 + \frac{[H]_0}{k_{mH}}\right) + \frac{1}{V} \quad (7)$$

where k_{mH} is a Michaelis-like constant for hydrogen ions. This equation has been applied[42] to fermenting yeast (Fig. 1.4) to determine relative affinities of various ions for the physiological potassium carrier, and to many transport systems particularly in epithelial tissue. Competitive

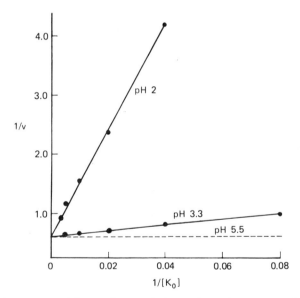

Figure 1.4 Competitive inhibition of K^+ uptake by H^+ during fermentation of yeast in media containing varying amounts of KCl, as illustrated by Lineweaver–Burk plot. From E. J. Conway, P. F. Duggan & R. P. Kernan, *Proc. Roy. Irish Acad.* **63B**, 93 (1963).

binding of hydrogen ions was probably responsible for the marked decline in potassium uptake in barley roots[43] when external pH was reduced from 6 to 4.

Kinetic studies[43] have revealed a dual pattern of potassium uptake by excised barley roots which may be attributed to different mechanisms or sites of accumulation. When the rate of ^{42}K absorption in the presence of 0.5 mM Ca was expressed as a function of $[K]_0$, this was in accordance with Michaelis–Menten kinetics over the $[K]_0$ range of 0.002 to 0.2 mM, reaching 90% of maximum uptake rate at 0.2 mM with a K_m value of 0.02 mM. When $[K]_0$ was then increased to 50 mM, the rate of uptake rose to a value much greater than that predictable from the kinetics over the lower concentration range. It was suggested that uptake over the lower and higher external concentration ranges occurred through different mechanisms. The lower range showed a much higher potassium affinity and was relatively insensitive to increase in external sodium and hydrogen ion concentration provided calcium was present. It was also insensitive to the nature of the external anion as between chloride or sulfate. The second mechanism, which made little contribution of potassium uptake at 1 mM K_0, had a much lower affinity for potassium and was inhibited by increases in sodium concentration. It was also inhibited when chloride was replaced by sulfate in the medium and by increase in the calcium concentration of the medium. Because of its fast response to changes in $[Ca^{2+}]_0$ and the fact that the range involved was comparable to that of soil water, it has been suggested that the response at the lower $[K]_0$ range reflects absorption at the plasmalemma, while the second range is believed to be effective at the tonoplast and concerned with osmoregulation. A simplified series model of transport across the two membranes has been criticized and a more complicated allosteric model based on multiphasic isotherms put forward.[44] Such a model would seem to rule out diffusion of potassium across plasmalemma and would negate the dual mechanism outlined above, but the matter is by no means settled yet, which is not surprising considering the complexity of the material under study.

Finally in regard to storage tissues,[45] electrolyte concentrations and membrane potentials were measured in beet root tissue in solution of KCl and of KHCO$_3$. In 0.2 mM KCl, E_m was -165 mV and was depolarized by 58 mV for a 10-fold change in $[K]_0$ indicating high permeability to this cation. Potassium here was close to electrochemical equilibrium across the membrane. However the substitution of bicarbonate for chloride caused a 70 mV hyperpolarization of the cell, associated apparently with induced active transport of protons out of the cells or bicarbonate in, the potassium apparently diffusing passively into the cells under the influence of the potential difference between E_m and E_K.

REFERENCES

1. J. G. Schilling, C. K. Unni & M. L. Bender, *Nature* (Lond.) **273**, 631 (1978).
2. H. Clarke, *The Data of Electrochemistry*, Bulletin 777 of the U.S. Geological Survey, 1924.
3. E. J. Conway, *Proc. Roy. Irish Acad.* **48B**, 161 (1943).
4. X. L. Pichon, *Nature* (Lond.) **267**, 765 (1977).
5. W. H. Twenhofel, *Principles of Sedimentation*, McGraw-Hill, New York & London, 1939.
6. A. I. Oparin, *The Chemical Origin of Life*, Charles C Thomas, Ill., 1964.
7. G. Knayse, in *Elements of Bacterial Cytology*, Cornstock, Ed., Ithaca, N.Y., 1951, p. 155.
8. W. Epstein & S. G. Schultz, in *Microbial Protoplasts, Spheroplasts and L-Forms*, L. B. Guze, Ed. Williams & Wilkins, Baltimore, Md., 1968, p. 186.
9. L. Wiklander, in *Potash Symposium*, International Potash Institute, Bern, Switzerland, 1954, p. 27.
10. M. Chevalier, *Potash Review* 16/53, International Potash Institute, Bern, Switzerland, 1971.
11. H. Jenny & R. Overstreet, *Proc. Nat. Acad. Sci. Wash.* **24**, 384 (1938).
12. S. B. Hendricks, R. A. Nelson & L. T. Alexander, *J. Am. Chem. Soc.* **62**, 1457 (1940).
13. N. Kamiya, in *Protoplasmalogia* VIII/3a, Springer-Verlag, Vienna, 1959.
14. L. N. Vorobiev, *Nature* (Lond.) **216**, 1325 (1967).
15. R. M. Spanswick & E. J. Williams, *J. Exp. Bot.* **15**, 193 (1964).
16. R. M. Spanswick, *Biochim. Biophys. Acta* **288**, 73 (1972).
17. K. Oda, *Sci. Rep. Tohoku Univ. Ser. IV Biol.* **27**, 159 (1961).
18. A. B. Hope & N. A. Walker, *Aust. J. Biol. Sci.* **14**, 26 (1961).
19. H. Kitasato, *J. Gen. Physiol.* **52**, 60 (1968).
20. O. F. Hutter & A. E. Warner *J. Physiol.* **189**, 403 (1967).
21. N. A. Walker & A. B. Hope, *Aust. J. Biol. Sci.* **22**, 1179 (1969).
22. A. L. Hodgkin & R. D. Keynes, *J. Physiol.* **128**, 61 (1955).
23. G. P. Findley, A. B. Hope & E. J. Williams, *Aust. J. Biol. Sci.* **22**, 1163 (1969).
24. R. W. Blount & B. H. Levendahl, *Acta Physiol. Scand.* **49**, 1 (1960).
25. J. S. Graves & J. Gutknect, *J. Gen. Physiol.* **67**, 579 (1976).
26. J. S. Graves & J. Gutknecht, *J. Membr. Biol.* **36**, 65 (1977).
27. C. L. Slayman, *J. Gen. Physiol.* **49**, 62 (1965).
28. S. G. Schultz & A. K. Solomon, *J. Gen. Physiol.* **45**, 355 (1961).
29. A. Rothstein, *Bacteriol. Rev.* **23**, 175 (1959).
30. C. L. Slayman & D. Gradmann, *Biophys. J.* **15**, 968 (1975).
31. D. Gradmann & C. L. Slayman, *J. Membr. Biol.* **23**, 181 (1975).

32. A. Seaston, G. Carr & A. A. Eddy, *Biochem. J.* **154**, 669 (1976).
33. M. Cockburn, P. Earnshaw & A. A. Eddy, *Biochem. J.* **146**, 705 (1975).
34. C. Slayman & E. E. Tatum, *Biochim. Biophys. Acta* **109**, 184 (1965).
35. D. J. F. Bowling & P. E. Weatherley, *J. Exp. Bot.* **16**, 732 (1965).
36. H. H. Ussing, *Acta. Physiol. Scand.* **19**, 43 (1949).
37. H. Linderholm, *Acta Physiol. Scand.* **27**, Suppl. 97, 1 (1952).
38. N. Higinbotham, B. Etherton & R. J. Foster, *Plant Physiol.* **42**, 37 (1967).
39. R. F. Davis & N. Higinbotham, *Plant Physiol.* **57**, 129 (1976).
40. E. Epstein & C. E. Hagen, *Plant. Physiol.* **27**, 457 (1952).
41. H. Lineweaver & D. Burk, *J. Am. Chem. Soc.* **56**, 658 (1934).
42. E. J. Conway, P. F. Duggan & R. P. Kernan, *Proc. Roy. Irish Acad.* **63B**, 93 (1963).
43. E. Epstein, *Nature* (Lond.) **212**, 1324 (1966).
44. A. D. M. Glass, *Plant Physiol.* **58**, 33 (1976).
45. R. J. Poole, *J. Gen. Physiol.* **49**, 551 (1966).

2
The Measurement of Cellular Potassium

The level of this cation in the cells of the body varies from tissue to tissue, and indeed from cell to cell, but its mean concentration in a particular tissue of the normal healthy individual remains remarkably constant throughout most of adult life. Where a significant fall occurs in its concentration, it is generally an indication of disease associated perhaps with hormonal deficiency, as in diabetes mellitus with acidosis or local ischemia.[1,2] It may also be caused by use of certain diuretics. From the clinical viewpoint, it is perhaps the variations in extracellular or plasma potassium concentrations which are most easily accessible to measurement and most detrimental in their effects on the body; but it has been recognized that accurate and reliable methods of assessing intracellular potassium and total body potassium are desirable as a diagnostic tool. The main methods that have been used to measure tissue potassium have been direct methods by analyses of tissue biopsies and indirect methods by determination of total body potassium through counting of the radioactivity of the naturally occurring isotope ^{40}K. A less satisfactory way of estimating total body potassium is a dilution method in which the artificial isotope ^{42}K is injected or taken by mouth, its specific activity at equilibrium distribution giving an indication of total exchangeable potassium in the body. All methods in use today have their advantages and disadvantages. It is proposed to review them and mention more recent methods of analysis of intracellular potassium concentration which may in time be adapted to a clinical setting.

The average male human body contains about 3200 m.mole of potassium dissolved in about 40 litres water. There is very little cation in adipose tissue. Therefore the female body, which has a higher percentage of this tissue, contains only about 2400 m.mole of potassium. Because of the variability of the amount of adipose tissue in man, it is more meaningful to express total body potassium in terms of fat-free body weight. A method of measurement of total body fat has been described[3] which makes use of the fat solubility of inert gases and is based on a dilution principle. The subject is placed in a closed respiration system containing traces of inert gas which then becomes absorbed by the body to an extent depending, at equilibrium, on the fat content of the subject. When a correction was made and potassium expressed on a fat-free tissue basis, the values in males and females were comparable. The fat-free body contains about 72% water in which most of the potassium is dissolved. This in turn may be divided into an extracellular fraction containing only about 70 m.mole K^+ and an intracellular fraction (55%) containing about 3000 mEq K^+. While total water has been measured by the dilution of administered tritiated water[4] and labeled antipyrine,[5] the extracellular fraction has been determined from the dilution of radiolabeled substances, such as inulin, which do not penetrate the cell membrane.[6] These are discussed later. As some of these substances are lost through the kidney before they reach equilibrium distribution throughout the extracellular space, infusion rather than a single injection is required. Intracellular water volume is obtained indirectly by subtracting the extracellular water volume, which of course includes plasma water, from total body water. The concentration of potassium in the cell water may then be calculated from the equation

$$[K]_i = \frac{TBK - ECF \times 4.5}{TBW - ECF} \text{ m.mole} \cdot \text{kg}^{-1} \text{ cell water} \qquad (8)$$

where TBK is the total body K in m.mole·kg^{-1} body weight, ECF is extracellular fluid volume in ml·g^{-1}, and TBW is total body water in ml·g^{-1}. The mean potassium concentration in extracellular fluid is taken to be 4.5 m.mole·kg^{-1} water here.

ANALYSIS OF TISSUE BIOPSIES

Since skeletal muscle makes up about 40% of body mass and is fairly accessible to biopsy, particularly by needle,[7] it has been the tissue of choice in the assessment of body potassium status in most cases. A fall in muscle content of this cation might be due either to a decrease in intracellular concentration, as seems to be the case in congestive heart failure, or to an increase in the extracellular relative to intracellular fluid volume of the tissue. While the patient's comfort dictates that as little tissue as possible be taken for analysis, when samples as small as 25 mg were assayed, their potassium concentrations varied greatly,[8] while samples of 150–200 mg did not show as wide a range of values.[9]

Because skeletal muscles are heterogeneous,[10,11] containing both pale phasic fibers with high potassium concentrations and red tonic fibers with lower potassium levels, it seemed possible that the variation might be due to this fact. Extensor digitorum longus muscles of rat, for example, contain potassium at a concentration of 173 m.mole·l^{-1} fiber water, compared with 150 m.mole·l^{-1} fiber water in soleus muscles. When a frequency distribution of human muscle biopsy values was plotted, however, there was no evidence of bimodality.[12] The cause of variability was probably elsewhere. It has been suggested[13] that the smaller the sample, the greater is its content of sodium chloride and water; it was therefore suggested that variable surface contamination with extracellular fluid might be responsible for the varying potassium content. It should also be mentioned that even in the case of isolated whole muscles of rat,[14] the ratio of dry to wet weight became very variable when muscle weight was below 50 mg, and there was a negative correlation between the volume of the extracellular fluid and muscle weight below this value. This relationship was not seen at 100 mg weight and over. It was essential therefore to know whether the wide variation in potassium content of the biopsy samples was real or an artefact produced by the smallness of the samples. From an extensive analysis of the interrelationships between electrolyte and water contents of small biopsies, it was concluded[8,15] that the variation was orderly in the following respects: (a) There was a positive correlation between the sodium and chloride in the samples and also between the water and the sum of potassium and sodium. (b) There was also a negative correlation between potassium and sodium except when these were expressed as amounts per kilogram dry weight of tissue; the correlation existed, for example, when fat-free dry weight (Fig. 2.1) was used. (c) In all cases the residual variability about these relationships was much less than the total variability. Taking advantage of the lesser stochastic variability, it was concluded that potassium analysis of a small biopsy afforded a reasonable index for whole muscle, provided changes in sodium chloride and water were also taken into account. The water content of the biopsy also appeared to give an accurate estimate of the water *in situ*. The conclusion also applied to biopsy samples from smooth and cardiac muscle as well as intestinal mucosa, brain, and connective

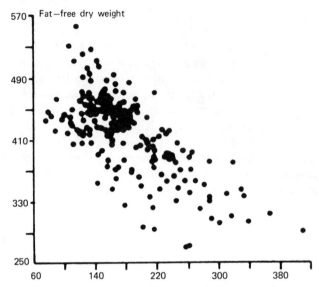

Figure 2.1 Relationship between sodium (abscissa) and potassium (ordinate) in material from nine muscles and 12 subjects (m.mole·kg^{-1} fat free dry wt.). From C. T. G. Flear, R. G. Carpenter & L. Florence, *J. Clin. Pathol.* **18**, 74 (1965).

tissue. However it was noted that while the potassium content of skeletal muscles appeared to be reduced in patients with congestive heart failure reflecting the fall in myocardial potassium, similar changes were not seen in brain tissue. Some other instances of localized potassium loss from tissues due to ischemia have been mentioned which were not manifest in skeletal muscle.

Regarding the analytical procedure, removal of visible fat and blood vessels from the muscle sample was evidently desirable, but occult fat was subsequently extracted and measured. Weighings of freshly removed frozen samples were carried out on a torsion balance graduated to 0.1 mg and potassium and sodium were analyzed by flame photometry after ashing. Chloride was determined amperometrically.

Finally by consideration of the sum of potassium and sodium concentrations in whole muscle and extracellular fluid it was shown that a distinction could be made between changes in muscle potassium content produced by variation of intracellular potassium and that produced by the shift of water between intracellular and extracellular compartments.

WHOLE BODY COUNTERS

The relative abundance of naturally occurring isotopes of potassium is 93.1% ^{39}K, 6.9% ^{41}K, and 0.118% ^{40}K. Other radioisotopes that can be produced by an atomic reactor include ^{37}K (half-life 1.2 sec), ^{38}K (7.7 min), ^{42}K (12.4 hr), ^{43}K (22 hr), ^{44}K (22 min), and ^{45}K (34 min). Because it is difficult to separate these, ^{40}K and ^{42}K have been predominantly used in the determination of body potassium. The former emits 11% gamma rays and 89% beta rays with a half-life of 1.3×10^9 years, being transmuted in the process to ^{40}Ca. The whole body counter has been used[16] to measure the natural radiation of ^{40}K and from this, total potassium has been calculated. Because of the weakness of the energy it has been necessary to take exceptional steps to minimize background radiation.

The scintillator is usually a sodium iodide crystal approximately 23 by 10 cm, coupled with a photomultiplier. Greatest sensitivity has been obtained with the patient in a chair tilted backward and with the scintillation crystal equidistant from chest and thighs.[17] The crystal requires a large angle of view, so local shielding is minimal. The counter is usually within a steel room. It has been found that a pre-1945 naval steel plate up to 20 cm thick with a 0.3 cm lining of aged lead is more satisfactory than modern steel because of its lower radiation level. Counting time varies from 200 to 1000 sec, yielding a body potassium value accurate and reproducible to 5% with a background count in the ^{40}K energy band (0.75–2.0 MeV) of 1.8×10^4 cpm. Special precautions had to be taken in calibrating the system because of the effect of geometry on the self-absorption of gamma radiation.

As control for the human subject, a phantom of polystyrene[18] or a polyethylene bottle[17] with the approximate geometry of the subject was used. This contained distilled water and was counted in the whole body counter for up to 1000 sec. A quantity of potassium chloride roughly equivalent to that of the body was then dissolved in the water and the phantom recounted. The increase in counts was due to the ^{40}K, which made up about 0.118% of the dissolved salt. This count could then be compared with that of the subject. However it was necessary to know also the efficiency of the counting of the ^{40}K. To determine this efficiency,[18] about 0.75 μCi ^{42}K was taken in a hypodermic syringe and placed in a standard position beside subject or phantom and counted. This isotope was then injected into the subject and allowed to equilibrate before recounting or was dissolved in the water contained in the phantom and again counted. A comparison of these various counts made it possible to assess the error due to self-absorption of radiation and hence the efficiency of counting, expressed in cpm·g^{-1} potassium.

A ratio of counting efficiency of ^{40}K over ^{42}K was also determined. The main problem with this method was the relatively long time required for adequate distribution of the ^{42}K throughout the body and the correction for loss in the urine. The phantom was usually an elliptical cylinder and therefore not strictly 4π in its geometry. In spite of this drawback it was considered possible that absorption of radiation could be allowed for in terms of the radial thickness of a cylindrical subject by the use of the term $(W/H)^{1/2}$, where W is the weight of the subject in kilograms and H is the height in centimeters. With a view to dispensing with the time-consuming injection of ^{42}K, the following formula was tested[18] which relates body build and ^{40}K counts with the total body potassium content TBK:

$$TBK = a \exp b(W/H)^{1/2} \times \text{net } ^{40}\text{K counts} \qquad (9)$$

The values of the constants a and b were found to be about 3.32 and 1.29, respectively, for 42 subjects, 17 of whom had metabolic disturbances associated with potassium deficiency.

In the course of this work regression equations were derived[19] for whole body potassium content, as measured with the whole body counter against total body weight on the one hand and against total body water measured by tritiated water dilution on the other. These equations were as follows:

$$TBK \text{ (g)} = (1.05 \pm 0.22)W \text{ (in kg)} + 48.5 \pm 15.5 \qquad \text{S.D.} \pm 12.4 \qquad (10)$$

and

$$TBK = (3.12 \pm 0.32)TBW \text{ (in l)} - (5.84 \pm 13.19) \qquad \text{S.D.} \pm 7.76 \qquad (11)$$

These results illustrate how correction for the fat contents of the body provides a more consistent relationship between the variables. The largest standard error value for any patient was $\pm 17\%$ on the basis of total body weight and $\pm 11.8\%$ on the basis of body water. The final conclusion of these studies was that only 16 min approximately were required to get statistically reliable and accurate results with a standard error of $\pm 4.7\%$ for potassium contents in the region of 100 g and over.

The importance of selecting the correct units in expressing results is shown in Table 2.1 of total body potassium measured in two cadavers of widely different builds, which were analyzed chemically after cremation.[20]

Table 2.1 Total Body Potassium of Cadavers[a]

Specimen	Total (m·mole)	m·mole·kg^{-1} Body Weight	m·mole·kg^{-1} Fat-Free Body Weight
Specimen 1: male, 47 yr old	2878	54.4	66.5
Specimen 2: male, 60 yr old	3543	48.6	66.6

[a]From Ref. 2.

The recognition of a body potassium deficiency, which in many cases denotes loss of intracellular potassium, depends on establishing the norm for each individual in relation to build and also the acceptable range of values about the norm. Total body counters as employed at present seem to be the most satisfactory methods available for detecting clinically significant loss of cellular and body potassium. However from the description given of the facilities required for the counting of ^{40}K, it should be evident that less expensive methods must be available. Such a method is the dilution of injected ^{42}K or the measurement of exchangeable body potassium,[21,22] K_e, which will now be considered briefly.

EXCHANGEABLE POTASSIUM MEASUREMENT

The naturally occurring isotope ^{41}K when bombarded with neutrons in a nuclear reactor is converted to ^{42}K; and as the former makes up only 6.9% of natural potassium, the specific activity, that is, the ratio of active to total K atoms in the irradiated sample, is relatively low. The short half-life of 12.4 hr is another limitation in its use for total body potassium determination. The determination of exchangeable potassium is carried out by injecting ^{42}K of known specific activity into the body or administering it by mouth and allowing it to equilibrate with the total potassium in the body. A sample of blood[23] or urine[24] is then analyzed, and from a comparison of specific activity before and after equilibration the total inactive pool of potassium with which the isotope has mixed may be calculated. Counting of the samples is a simple matter in this case, but error arises from the inadequate mixing or equilibration of the isotope. As a result values found for exchangeable potassium are usually about 90–95% of those obtained by total body counting. It would seem desirable therefore to seek evidence of equilibrium by examination of serial samples starting at about 24 hr after injection.

It is necessary that the specific activity of the injected ^{42}K should be as high as possible to allow for the large dilution.[25] About 60×10^6 cpm in 0.15 m.mole·l^{-1} potassium chloride has been recommended as a suitable level for injection. Electromagnetic concentration of ^{41}K in naturally occurring potassium has made the production of ^{42}K of higher specific activity possible.

After its injection ^{42}K is rapidly lost from the blood, so that 90% is removed from the circulation within 1 min and only 5% remains after the blood has circulated through the body three times.[26] It becomes most concentrated in renal tissue at first, and about 20% of the label there is subsequently lost in the urine. Lungs and intestine have high turnover rates while muscle and erythrocytes have relatively low turnover rates. The potassium exchange in muscle is not complete until nearly 12 hr after injection. The relative counts per unit weight of potassium of injected sample and that of urine or blood after mixing should give a measure of the exchangeable potassium throughout the body, but whether exchangeable potassium and total body potassium are identical is questionable. When K_e was expressed as a percentage of total body potassium measured by whole body counter in normal subjects, the values[27] ranged from 97% down to 94.7% (based on plasma analysis) and 91.5% (urine analysis)[28] and even as low as 85%.[29] The reason for this variation is probably that equilibrium distribution had not been reached. It has been suggested[29,30] that (a) 40 hr is required for equilibrium to be established and that 24 hr used previously is not adequate, (b) complete mixing should be investigated perhaps by taking several samples toward the end of the dilution period, and (c) that the final value should be based on at least four measurements.

Total body potassium and exchangeable potassium are not necessarily indicators of intracellular concentration but may be due to changes in the relative proportions of cellular to extracellular phases (Fig. 2.2). It has been possible in some cases[15] to assess the contribution of changes in water distribution between these compartments on tissue electrolyte con-

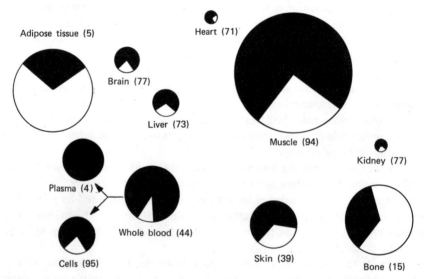

Figure 2.2 Potassium concentrations in various organs and tissues of the human body in m.mole·kg^{-1} tissue (m.mole·l^{-1} in fluids). Areas of circles are proportional to weights of tissues, areas in white are proportional to dry weights. From R. P. Kernan, *Cell K*, Butterworths, 1965.

centrations. For example, a negative correlation may be expected between the sodium concentration in tissue water and the sum of the sodium plus potassium concentration in this fluid when these change as a result of water movement alone. The accurate determination of intracellular potassium concentration from total tissue or body potassium requires a knowledge of water distribution between the two phases and therefore an accurate measurement of extracellular water. Unfortunately the latter cannot be accurately measured because indicators used for this purpose cannot reach the transcellular fluid.

The properties of the ideal probe for measuring *ECF* may be summarized as follows: (a) The probe should become rapidly and homogeneously distributed throughout the ECF; (b) it should not enter the cells; (c) it should be nontoxic; (d) it should not be metabolically transformed or bound by proteins; and (e) its elimination from the body should be easily measurable. The probes that have been used for this purpose include inulin,[31] which diffuses slowly, sulfate,[32,33] thiosulfate,[34] thiocyanate,[35] mannitol,[36] EDTA,[37] and sucrose, none of which meets all these requirements. Another indicator which appeared[38] to move with plasma proteins during electrophoresis and therefore to be bound by them was ferrocyanide. More recently it has been found[39] that 98.9 ± 2.1% of the ferrocyanide may be recovered from the urine after injection into dogs and that its partition between lymph water and plasma water was 0.99 ± 0.02, which would rule out significant binding. This substance would seem to fulfill many of the requirements of a satisfactory probe. While the methods already described may provide a reasonable estimate of the body's electrolyte status in the normal individual, particularly in respect to potassium ions, unfortunately the same cannot always be said of the patient with disorders of electrolyte metabolism. Hopefully in the future the direct measurement of intracellular potassium of cells *in vivo* may be possible and this would remove the uncertainty associated with water distribution.

Direct Measurement of Intracellular Potassium by Means of Ion Selective Microelectrodes: The two principal types of potassium selective microelectrodes which have been used for the measurement of the activity of this ion in animal cells *in vivo* and *in vitro* have been those prepared from potassium-selective glass[40] and those made from common borosilicate glass but filled with potassium-selective liquid ion exchanger.[41] The sensitivity of glass to alkali metal ions over hydrogen ions depends on the presence of trivalent oxides in the glass. If their concentration exceeds 10 moles %, the glass exhibits Nernstian response to sodium in the range of 10^{-3} to 1 M within the pH range of 6 to 10. Such glass responds to potassium also, but to a lesser extent.

By varying the proportions of sodium to aluminum in the glass, it has been possible to vary the selectivity coefficient to sodium over potassium over a fairly wide range. At best such glass is about 10 times more selective for potassium than for sodium; and at the other end of the scale, NAS_{27-4} glass (27 mole % Na_2O, 4 mole % Al_2O_3, 0.69 mole % SiO_2) shows a selectivity for sodium about 200 times greater than for potassium. The latter glass is also more durable than the former. In spite of the low selectivity of such glass for potassium, it has been possible by making use of glasses with a wide range of relative selectivity to make quite accurate measurements of the activity of potassium within the fibers of skeletal muscles.[42-44] Thus by the simultaneous measurement of concentrations by the indirect chemical method outlined above, activity coefficients have been determined for the potassium ions within the fibers. It should be mentioned however that the construction of microelectrodes using the ion-selective glass is by no means an easy matter since the glass needs to be insulated by painting with shellac or polystyrene paint to within a few microns of its 0.5 μm diameter tip before it can be used. Recessed tip microelectrodes of this glass[45] (Fig. 2.3) have obvious advantages over insulation by paint.

Microelectrodes containing the Corning liquid-ion exchanger have a number of clear advantages over the potassium-selective glass in the measurement of intracellular potassium activity of tissues. Firstly the selectivity coefficient k_{KNa} of these microelectrodes indicate that they may be up to 70 times more selective for potassium than for sodium. This property of the electrode is measured by comparing the potential of the potassium microelectrode with respect to the conventional 3 M KCl-filled microelectrode in solutions of KCl and NaCl of the same molarity, the coefficient calculated from the equation

$$k_{KNa} = e^{\frac{F(E_{Na} - E_K)}{RT}} \qquad (12)$$

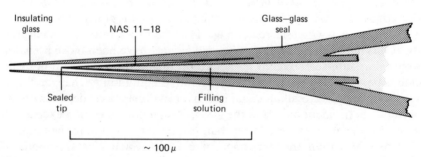

Figure 2.3 Diagram showing construction of recessed-tip Na^+-sensitive microelectrode. From R. C. Thomas, *J. Physiol.* **220**, 55 (1972).

where E_{Na} and E_K are the measured potentials. The potassium electrode is prepared by drawing ordinary borosilicate glass to a tip diameter of about 0.5 μm and then making the inside of the tip hydrophobic for about 150–200 μm by treatment with trimethyl chlorosilane or a similar substance. The liquid ion exchangers are solutions of organic electrolytes in water-immiscible solvents, and these cannot be maintained within the tip if the surface of the glass is hydrophilic. The ion exchange material of the potassium-selective electrode is based on K^+-tetra(p-chlorophenyl) borate or tetra(p-phenoxyphenyl) borate dissolved in 3-o-nitroxylene.[46]

Nicolsky[47] derived the potential of an ion-selective electrode immersed in a solution containing two univalent cations to which it was sensitive to different degrees. This may be expressed as

$$E = E_0 + \frac{RT}{F} \ln(a_i + k_{ij} a_j) \tag{13}$$

where a_i is the activity of the preferred cation E_0, a constant; and k_{ij} is the selectivity coefficient of the electrode. With the aid of this equation and with both electrodes introduced into the same cell or muscle fiber, the intracellular activity of potassium ions may be readily determined. To facilitate such measurements, double barreled electrodes with tip diameter of less than 1 μm may be prepared[48] in which one barrel is a potassium electrode and the other, the conventional electrode. The ion-selective microelectrode is first calibrated in solution of known cation activity. It may then be introduced into the cell, whereupon the potential change ΔE observed will be the difference between the membrane potential and the change in cation activity detected by the ion selective electrode on entering the fiber. This may be expressed as follows:

$$\Delta E = E_m - E_{KNa} = E_m - \frac{RT}{F} \ln \left[\frac{a_K^o + k_{KNa} a_{Na}^o}{a_K^i + k_{KNa} a_{Na}^i} \right] \tag{14}$$

If the mean value of ΔE is calculated from measurements on about 40 cells or fibers and the mean membrane potential of the same fibers is subtracted from this, we have the term E_{KNa}, from which the intracellular potassium activity may be readily determined from eq. (14) above. It should be mentioned that in the case of potassium-selective electrodes filled with ion exchanger in which the term k_{KNa} may be as small as 0.012 and intracellular sodium activity a_{Na}^i of skeletal muscle fibers as little as 6 m.mole at rest, the term $k_{KNa} a_{Na}^i$ may be omitted from the equation without significant error.

The main limitation in the use of the ion-selective liquid exchanger microelectrode is its high electrical impedance, which is of the order of 10^9 ohms for a tip diameter of about 0.5 μm. It is necessary therefore to have a high input impedance in the amplifier to which the electrodes are coupled. This may be achieved by a differential amplifier using field effect transistors (FET) with impedances of up to 10^{14} Ω.

The potassium ionophore valinomycin, first obtained from cultures of *Streptomyces fulvissimus*,[49] is a macrocyclic ligand with a 36-membered ring. It undergoes a conformational change complexing potassium within its inwardly directed polar groups while its lipophilic side chains are directed outward.[50,51] The coordination number of potassium within the complex is 6, and the diameter of the cavity is 2.6–3.2 Å, compared with an ion crystal radius of 1.33 Å for potassium. The formation of complexes with macrocyclic compounds is brought about by an interaction between the desolvated ion with the dipoles inside the ligand cavity. Since the binding of potassium ions by valinomycin is reversible, particularly at the interfaces of a lipid membrane, it might be concluded that this substance would make excellent potassium-selective electrodes. Electrodes with tips containing valinomycin dissolved in diphenyl ether had a k_{KNa} value of 2.6×10^{-5}, making them virtually unresponsive to sodium in the presence of potassium. However because of their exceptionally high impedance they have not been used in microelectrodes, although a modified form containing also tetraphenyl borate and other substances in a PVC base has been used for the measurement of potassium in the circulation in cardiac bypass surgery.[52] In this case of course a microcapillary electrode was not required.

It is not inconceivable, however, that with technical developments in this field, suitable intracellular potassium electrodes may become available for clinical use during and after surgery.

REFERENCES

1. J. H. Bland, *Clinical Metabolism of Body Water and Electrolytes*, W. B. Saunders & Co., Philadelphia 1963.
2. C. T. G. Flear, in *Electrolytes and Cardiovascular Diseases*, Vol. 2, E. Bajusz, Ed. S. Karger, N.Y., 1966, p. 357.
3. G. T. Lesser, W. Perl & J. Murray Steele, *J. Clin. Invest.* **39**, 1791 (1960).
4. T. C. Prentice, W. Siri, N. I. Berlin, G. M. Hyde, R. J. Parsons, E. E. Joiner & J. H. Lawrence, *J. Clin. Invest.* **31**, 412 (1952).
5. R. J. Soberman, B. B. Brodie, B. B. Levy, J. Axelrod, V. Hollander & J. M. Steele, *J. Biol. Chem.* **179**, 31 (1949).
6. I. S. Edelman & J. Lieberman, *Am. J. Med.* **27**, 256 (1959).
7. J. Bergström, *Scand. J. Clin. Lab. Invest.* **14**, Suppl. 68 (1962).
8. C. T. G. Flear, R. G. Carpenter & I. Florence, *J. Clin. Pathol.* **18**, 74 (1965).

References

9. F. P. Muldowney, in *Compartments, Pools and Spaces in Medical Physiology*, P. E. Bergner, Ed. C. C. Lushbough, 1967, p. 137.
10. F. A. Streter & G. Woo, *Am. J. Physiol.* **205**, 1290 (1963).
11. J. F. Y. Hoh & B. Salafsky, *J. Physiol.* **216**, 171 (1971).
12. C. T. G. Flear, I. Florence & J. A. Williams, *J. Clin. Pathol.* **21**, 555 (1968).
13. C. L. Johnson, *Clin. Sci.* **28**, 57 (1965).
14. N. Kobayashi & K. Yonemura, *Jpn. J. Physiol.* **17**, 698 (1967).
15. C. T. G. Flear, *Ann. N.Y. Acad. Sci.* **156**, 421 (1969).
16. P. Delwaide, W. G. Verly, J. Colard & R. Boulenger, *Health Phys.* **9**, 147 (1963).
17. G. R. Meneely, C. O. Ball, J. L. Ferguson, D. O. Payne, A. R. Lorimer, R. L. Weiland & H. L. Wolf. *Circ. Res.* **11**, 539 (1962).
18. D. Hughes & R. E. Williams. *Clin. Sci.* **32**, 495 (1967).
19. D. Hughes, R. E. Williams & A. H. Smith, *Clin. Sci.* **32**, 503 (1967).
20. G. B. Forbes & A. M. Lewis, *J. Clin. Invest.* **35**, 596 (1956).
21. I. S. Edelman, J. M. Olney, A. H. James & L. Brooks; *Science* **115**, 447 (1952).
22. F. D. Moore, J. D. McMurrey, H. V. Parker & J. Caryl Magnus, *Metabolism* **5**, 447 (1956).
23. L. Corsa, J. M. Olney, R. W. Steenburg, M. R. Ball & F. D. Moore, *J. Clin. Invest.* **29**, 1280 (1952).
24. F. P. Muldowney, J. J. Haxhe, A. W. Marczynska & F. D. Moore, *J. Lab. Clin. Med.* **56**, 127, (1960).
25. C. L. Comar, in *Radioisotopes in Biology and Agriculture*, McGraw-Hill, New York, 1955.
26. W. G. Walker & W. S. Wilde *Am. J. Physiol.* **170**, 401 (1952).
27. J. Rundo & U. Sagild, *Nature* (Lond.) **175**, 774 (1955).
28. L. Surveyor & D. Hughes, *J. Lab. clin. Med.* **71**, 464 (1968).
29. S. A. Threefoot, *Progr. Cardiovasc. Dis.* **5**, 32 (1962).
30. C. T. G. Flear, in *Compartments, Pools and Spaces in Medical Physiology*, P. E. Bergner & E. S. Anderson, Eds., U.S. Atomic Energy Commission Publication, 1967, p. 53.
31. G. Nichols, Jr., N. Nichols, W. B. Weil & W. M. Wallace, *J. Clin. Invest.* **32**, 1299 (1953).
32. D. D. Dziewatkowski, *J. Biol. Chem.* **178**, 389 (1949).
33. M. Walser, D. W. Seldin & A. Grollman. *J. Clin. Invest.* **32**, 299 (1953).
34. A. Gilman, S. F. Philip, E. S. Koelle, *Am. J. Physiol.* **146**, 348 (1943).
35. A. W. Winkler, J. R. Elkington & A. J. Eisenman, *Am. J. Physiol.* **139**, 239 (1943).
36. J. R. Elkington, *J. Clin. Invest.* **27**, 1088 (1947).

37. H. Foreman, M. Vier & M. Magee *J. Biol. Chem.* **203**, 1045 (1953).
38. C. R. Kleeman & F. H. Epstein, *Proc. Soc. Exp. Biol.* **93**, 228 (1956).
39. J. Zweens, H. Frankena, P. Rispens & W. G. Zijlatra, *Pflügers Arch.* **357**, 275 (1975).
40. G. Eisenman, in *Glass Electrodes for Hydrogen and Other Ions*, G. Eisenman Ed., Marcel Dekker, New York, 1967, p. 268.
41. J. L. Walker, Jr., *Anal. Chem.* **43**, 89A (1971).
42. A. A. Lev, *Nature* (Lond.) **201**, 1132 (1971).
43. P. G. Kostyuk, Z. A. Sorokina & Y. D. Kholodova, in *Glass Microelectrodes*, M. Lavallee, O. F. Schanne & N. C. Hebert, Eds., Wiley, New York, 1969, p. 322.
44. W. McD. Armstrong & C. O. Lee, *Science* **171**, 413 (1971).
45. R. C. Thomas, *J. Physiol.* (Lond.) **210**, 82P (1970).
46. G. Baum & M. Lynn, *Anal. Chim. Acta* **65**, 393 (1973).
47. B. P. Nicolsky, *Acta Physicochim. U.S.S.R.* **7**, 597 (1937).
48. R. N. Khuri, J. J. Hajjar & S. K. Agulian, *J. Appl. Physiol.* **32**, 419 (1972).
49. H. Brockman & G. Schmidt-Kestner, *Chem. Ber.* **88**, 57 (1955).
50. B. C. Pressman, *Ann. Rev. Biochem.* **45**, 502 (1976).
51. J. Koryta, in *Ion Selective Electrodes*, Cambridge University Press, London, 1975, p. 138.
52. D. M. Band. J. Kratochvil & T. Treasure. *J. Physiol.* **265**, 5P (1976).

3
Potassium Equilibrium and the Resting Cell

The property of animal and vegetable cells, with the exception of red blood cells of some herbivora,[1] of accumulating potassium ions selectively while virtually excluding sodium has been attributed on the one hand to the properties of protein and water in the bulk phase of the cell and on the other hand to the properties of the cell membrane. According to the former view, as exemplified in the association–induction hypothesis of Ling,[2,3] potassium ions are selectively adsorbed on β- and γ-carboxyl groups of cellular protein, to which the bulk of cell water is also polarized and oriented in a multilayered state, the so-called structured water intermediate between water and ice. To quote Ling, "The living protoplasm of protein, ion and water represent a three-dimensional cooperative assembly under the control of certain cardinal adsorbents, of which ATP is a prime example." The role of ATP is not seen here as providing a package of energy in chemical bonds that can be used by ion pumps but rather as adsorbent, producing a change in electron distribution in the backbone peptide groups, thereby maintaining the protein–ion–water system at a higher energy state. Dephosphorylation of ATP is then associated with a reduction in the energy state of the whole assembly and presumably with loss of the specific adsorption of potassium.

The main weakness of this hypothesis is probably that it fails to account for the electrical properties of cells[4-6] that are the basis of excitability. In Ling's view the cell membrane has no special function in determining potassium accumulation and the selective permeability of the cell.

The membrane theory, on the other hand, postulates[7,8] the presence of ion pumps for the extrusion of such cations as sodium, calcium, and protons across the cell membrane, the presence in this structure of ion-selective channels regulating the passive movement of ions into and out of the cell, and a Donnan equilibrium with respect to permeable ions between cytoplasm and external medium. Active transport of ions has been demonstrated in red cell "ghosts" from which more than 90% of proteins and hemoglobin had been removed.[9] Such cells accumulate potassium without an apparent requirement for binding to proteins within the cell; the energy for the process comes from the hydrolysis of ATP. The only potassium binding evident in such cells is that to the integral proteins of the membrane which are considered to be part of the mechanism of the $Na^+ + K^+$–ATP-ase cation transport system.

ASSOCIATION–INDUCTION HYPOTHESIS

Cell membranes are leaky to ions to varying degrees, so that even in the so-called resting state they tend to lose potassium in exchange for sodium taken up. This sodium must be continually pumped out again with the use of metabolic energy. One of the earliest contentions of supporters of the "association–induction hypothesis" was that the energy available from metabolism was insufficient to account for the maintenance of ionic gradients between muscles and their surrounding media after they had been treated with iodoacetate and anoxia at 0°C, conditions calculated to deprive the cells of energy from both aerobic and anaerobic sources. To determine the energy requirements here, it was necessary to know the rate at which sodium leaked into the cell or was pumped out; this rate was usually measured with radioisotopes. However by this method it is possible to overestimate net movement of sodium across the membrane. For example, when skeletal muscles were labeled with ^{22}Na *in vivo* and the freshly isolated muscles immersed in inactive Ringer solution, it was shown[10] that the apparent efflux, as measured by loss of label from the muscles, could be accounted for almost completely by exchange of labeled for unlabeled sodium with negligible net efflux of this cation. The process of exchange diffusion[11] can therefore lead to a gross overestimation of the energy needed for active transport of ions.

More recently[12] the energy required for sodium and potassium transport in rat soleus muscle fibers has been estimated to be about 5–6% of the basal metabolic energy of the muscles. The energy for active transport in this case was taken as that part of total energy consumption which was eliminated by addition of ouabain, which inhibits the pump. The efficiency of the pump under these conditions appeared to be about 34%. As much as 56% of total energy turnover was found[13] to be required for net transport of sodium and potassium in the case of sodium-enriched muscles, which extruded up to 20 m.mole Na·kg^{-1} muscle water over a 2 hr period where this was based on oxygen consumption in control and transporting muscles.

In *Escherichia coli* the rate of hydrolysis of ATP was assessed from the uptake of labeled phosphate from the external medium; on this basis it was concluded[14] that the energy turnover was less than that needed for potassium accumulation by the cells. This was used as an argument in support of the association–induction hypothesis. These studies have been criticized since[15,16] on the grounds that the large intracellular pool of phosphate had not been taken into account. One might also add that the evidence[17,18] that proton gradients between cytoplasm and external media can energize solvent transport also lessens the force of the argument. Subsequent isolation[19] of a potassium-binding protein (MW 40,000) from the supernatant of homogenized cells has been taken as evidence of internal sequestration of this cation as a mechanism of accumulation. The nature and origin of this protein and its selectivity with respect to sodium have yet to be clarified.

The possibility of selective binding of potassium within skeletal muscle fibers has been examined by the measurement[20] of comparative mobilities of a number of ions and molecules within the sarcoplasm. The relative diffusion rates of ^{42}K, ^{36}Cl, and ^{14}C-sorbitol along the longitudinal axis of single fibers placed within glass capillaries were determined over a period of 24 hr in solutions of different pH values. The diffusion coefficient of potassium with respect to the other labeled substances was found to decrease with increasing pH, and this was taken as an indication of interaction between this cation and charges on internal proteins. The diffusion rate for chloride increased, while that of ^{14}C-sorbitol remained constant under these conditions. The principles of diffusion in polyelectrolyte solutions were applied to the data,[21] and the fraction of ions that appeared to be bound, called the binding fraction f_K or f_{Cl}, was calculated. The value of f_K increased from 0 to 0.13 as pH was increased from 5.2 to 10, with a reciprocal change in f_{Cl}.

While some interaction between ions and proteins was indicated, albeit over a wide pH range, a comparison with earlier work[22] suggested a selectivity for binding of sodium to potassium of 3.3:1; the total binding at pH 7.4 only amounted to about 70 m.mole·kg^{-1} dry weight of muscle, which was equivalent to about 14 m.mole·kg^{-1} wet weight of tissue. Earlier studies[23] of the relative diffusion rates of potassium, sodium, and sulfate ions of ATP and sorbitol in skinned skeletal muscle fibers indicated that chemical interaction with fixed charges within the cell did not influence their mobility but their diffusion coefficients within the fibers appeared to be only about half the magnitude expected of ions in free solution. It was suggested therefore that the retardation might have been due to "obstruction of structure within the cell (tortuosity factor) and increased viscosity of the cytoplasm." Recently the relative mobilities of ions in what has been termed "an effectively membraneless open ended cell" (EMOC) have been measured.[24] This preparation was frog sartorius muscle that had been centrifuged to remove interstitial fluid and one end amputated and immersed in Ringer fluid containing ^{22}Na and ^{42}K. The intact end of the muscle was passed through a gasket, intended to constrict further diffusion through interstitial fluid thereby ensuring that the main diffusion pathway would be intracellular. The intact portion of the preparation above the gasket was suspended in air. It was suggested that in this preparation the remaining anatomically intact membranes had been made nonfunctional by removal of interstitial fluid, which was a sink for outward pumping and a source for inward pumping of ions.

As expected, the preparation lost potassium and gained sodium at the damaged end, but the point of greatest interest was that the intact portion of the muscle eventually accumulated about three times as much ^{42}K as was contained in the Ringer solution but had less ^{22}Na than this fluid. This of course did not imply that the specific activity of potassium in the muscle was equal to or greater than that in the bathing fluid. The preferential uptake of ^{42}K over ^{22}Na was absent however when ouabain was included in the bathing fluid. The results were interpreted as favoring the association–induction hypothesis.

The validity of this interpretation depends on whether the muscle fiber can be regarded as a single compartment system without internal membranes involved in sodium–potassium exchange or whether there may be appreciable extracellular fluid still present in the preparation. Evidence from flux measurements using radioactive ions,[11,25] from electronmicroscopic studies of muscles in hypotonic and hypertonic media,[26] and activity measurements with ion-selective microelectrodes[27] all lead to the conclusion that there exists a significant extracellular compartment other than interstitial fluid which may include the sarcoplasmic reticulum or sarcotubular system (see Plates 1–2). Centrifugation of muscles reduced the mean weight by 9.3%; that is, less than the 19.2% found by ^{14}C-inulin[28] which is probably a truer estimate of the total extracellular space in frog muscle including the sarcotubular system. The presence of such intrafiber

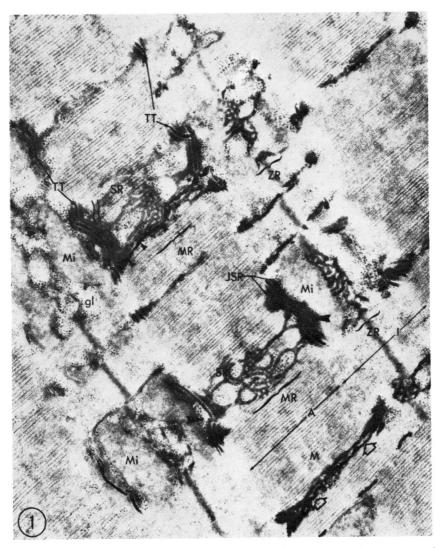

Plate 1 The transverse tubules (TT) located at the A-I junction, are filled with lanthanum and are generally darker than adjacent sarcoplasmic reticulum (SR), which is also filled with lanthanum. A longitudinal tubule (small arrow heads) is probably a longitudinal extension of the TT. The similarity in densities between the longitudinal tubule and junctional SR (JSR) (asterisk) gives the impression that the two systems are connected, but in stereo projection the former is out of phase with adjacent SR. The free SR is distributed in two tubular networks or retes overlying the M line (MR) and Z disk (ZR) regions of the sarcomere. Fenestrations occur in both retes and in the JSR. The latter fenestrations are generally smaller than the rete fenestrations and are aligned in rows parallel to the long axis of the coupling. A double row of fenestrations (curved arrows) is seen in the JSR on the upper left. The continuity of the SR is evidenced by the areas were JSR is continuous around the TT (large arrowheads). Mitochondria (M_i) and lightly stained glycogen (gl) are labeled. × 27,000. From R. A. Waugh, T. L. Spray & J. R. Sommer, *J. Cell Biol.* **59**, 254 (1973).

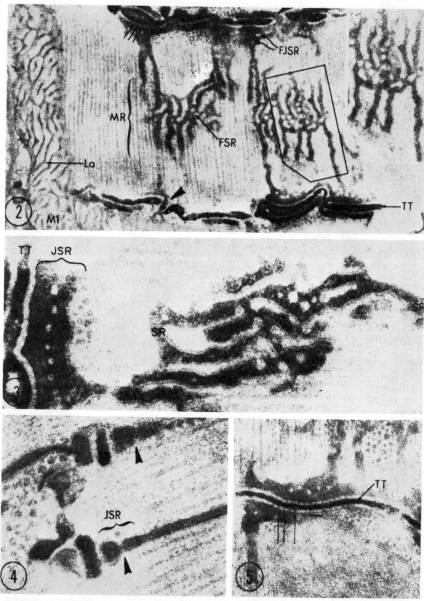

Plate 2 a. The tubular nature of SR rete overlying M line region (MR) is shown. Fenestrations of both SR (FSR) and junctional SR (FJSR) are seen. The FJSR (curved arrows) are aligned in rows, and in one location a single fenestration of a second row is seen. A mitochondrion (Mi) near the cell surface shows lanthanum (La) within the outer mitochondrial compartment. Inset is an enlarged SR rete. The lucent space separating SR membranes and La-filled space is apparent, as are the membrane boundaries of the fenestrations. Several areas are seen (large arrowheads) where junctional SR is continuous around the transverse tubules. ×53,000 (inset ×75,000).

compartmentalization might also explain the low diffusion coefficients that have been attributed to the tortuosity within the fibers.[23]

A critical question in relation to Ling's hypothesis is the relative magnitudes of the activity coefficients for sodium and potassium within the cell, because when an ion is bound its activity and activity coefficient should be reduced below those of the ions in free solution at a similar concentration. Therefore on the basis of this hypothesis the activity coefficient for potassium should be less than that of sodium. Intracellular activities of these cations have been measured[29] by ion-selective microelectrodes made of Corning glass in single immature oocytes of toad, which were about 700–900 μm in diameter. The oocytes were then analyzed by flame photometry to determine cation concentrations, after their water content had been estimated by drying. The mean potassium activity a_K of cell water was found to be 82 m.mole·l^{-1}, compared with a [K]$_i$ concentration of 113 m.mole·l^{-1} cell water, giving an activity coefficient γ_K of 82/113, or 0.73, which was close to that of a similar concentration of potassium in free solution. In the case of sodium, on the other hand, a_{Na} was 9.3 m.mole·l^{-1} and [Na]$_i$ was 25.8 m.mole·l^{-1} cell water, yielding 0.36 for γ_{Na} and indicating that about half of the sodium within the cell was unable to react with the microelectrode. Similar results have been obtained in nerve[30] and muscle[31] fibers. More recently similar measurements[32] were made in mature oocytes of frog where γ_{Na} turned out to be as low as 0.08 ± 0.02 and γ_K was as high as 1.15 ± 0.03. These unlikely results have been attributed to the development of subcellular compartmentalization of these cations.

Plant cells provide an instructive medium in which to test Ling's hypothesis, since they contain both the protein-rich cytoplasm and the vacuolar fluid, which contains little or no protein, and these have been examined separately. Potassium activities have been measured[33] in these fluids in the freshwater alga *Chara australis* and in the marine algae *Griffithsia* sp. In the former, a_K was about 2.3 times greater in the cyto-

---

b. The membrane boundaries of the fenestrations within JSR and SR. The 20–30 Å lucent space beneath SR membrane is continuous from free to junctional SR and includes the portion of the SR which faces the TT. A similar lucent line is present adjacent to La-filled TT. Note particulate substructure of SR in both locations.

c. Two triads composed of JSR flanking central TT are shown in cross section. The right-hand JSR appears to be sectioned at the level of the fenestration (arrowhead). This is also apparent in the lower triad. The complete gap through JSR is not seen because of section thickness in relation to small size of fenestration. ×146,000.

d. En face view of triad. Fenestrations are present on both sides of TT. In places periodic or junctional processes (arrows) appear to completely span gap between JSR and TT. ×115,000. From R. A. Waugh, T. L. Spray & J. R. Sommer, *J. Cell Biol.* **59**, 254 (1973).

plasm than in the vacuole, but in the latter it was 2.2 times greater in the vacuole than in the cytoplasm. The ratio of a_K in cytoplasm of marine algae to that of freshwater algae was about 1.4:1, while the respective ratio for the vacuoles was 7.3:1. It is significant that the vacuole, which contained little or no potassium-binding material, showed a greater range of potassium activity than did the cytoplasm.

THE MEMBRANE THEORY

According to this theory it is the semipermeability of the membrane surrounding the cytoplasm which determines the composition of the latter in relation to that of the external medium. It was suggested,[34] for example, that the membrane contained water-filled pores through which ions could pass only if their hydrated diameters were less than that of the pores. While the low permeability of the membrane for sodium with respect to potassium could be explained on this basis, it was not possible to account for the low permeability of caesium ions, which in their hydrated state should have a diameter similar to that of potassium.

This problem was subsequently solved[35] by the assumption that the naked ion radius had to correspond exactly to that of the mean pore radius for the ion to be permeable. The all-or-none element introduced by this concept was avoided by the dispersion in pore radii illustrated in Fig. 3.1. The idea of charged permeability channels favoring, for example, the

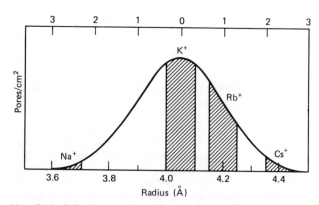

Figure 3.1 Hypothetical distribution of membrane pores with respect to size. Shaded areas represent the number of pores in the distribution that fit an ion of a particular size to within ± 0.05 Å. Ordinate is number of pores per unit area of membrane and abscissa is radius of the ion (crystal radius + 2.72 Å). At top, standard deviation is shown. From L. J. Mullins, *J. Gen. Physiol.* **42**, 817 (1959).

passage of an anion over a cation of similar size was later developed.[36] At one time it was believed that the membrane was impermeable to chloride[37] as well as to sodium, but this was subsequently disproved when Boyle and Conway[34] showed that both potassium and chloride were distributed between the cell water and external medium in a Donnan equilibrium. This relationship was expressed as

$$[K]_i \times [Cl]_i = [K]_0 \times [Cl]_0 \qquad (15)$$

where the activity coefficients of the ions were assumed to be similar inside and outside the cell. Because of the presence of nondiffusible molecules within the cell with net negative charge, the preferential accumulation of cations over anions by the cells was said to take place. However if the membrane was permeable to both sodium and potassium ions, the effect of the nondiffusible anion in a simple Donnan system would be to produce a higher internal osmotic pressure than that outside the cell, which could not be tolerated where the membrane was distensible.

It was therefore suggested that the presence of a nondiffusible cation outside the cell in the form of sodium ions would counteract the osmotic disequilibrium and prevent swelling of the cell. Conway's application of the Donnan equilibrium to the distribution of permeable ions across the cell membrane was tested by changing the composition of the fluid bathing isolated frog muscles and measuring the resulting changes in muscle weight, intracellular ion concentration, and injury potentials of the muscles. Since the membrane is impermeable to sodium ions, these along with chloride and bicarbonate are mainly responsible for the maintenance of the external osmotic pressure. When the external potassium concentration, which in frog Ringer was about 2.5 mM, was increased above 10 mM by the addition of solid potassium chloride to the Ringer, so that the product $[K]_0 \times [Cl]_0$ was increased, a net movement of these ions into the muscle fibers took place so as to restore the Donnan equilibrium. This uptake could sometimes be detected as a change in weight of the muscle, which first lost water to the hypertonic bathing fluid and then regained it along with the electrolytes. This experiment also demonstrated the permeability of the membrane to chloride. When, on the other hand, $[K]_0$ was increased by substitution for sodium in the Ringer, restoration of the Donnan equilibrium resulted in the external medium becoming hypotonic and in water uptake causing an increase in muscle weight. In subsequent experiments, $[K]_0$ was increased by substitution for sodium in such a way as to keep the product $[K]_0 \times [Cl]_0$ constant, that is, by replacing external chloride in stages by an impermeant anion. Under these conditions $[K]_i$ remained constant. These simple experiments yielded results which indicated that the Donnan equilibrium was indeed applicable to frog muscle, but the measurements of injury potential left something to be desired.

If the potassium and chloride ions are passively distributed across the membrane as suggested, then the inside of the cell should be negatively charged and the relationship between ionic gradients and membrane potential expressed by the relationship

$$\frac{[K]_o}{[K]_i} = \frac{[Cl]_i}{[Cl]_o} = e^{EF/RT} \qquad (16)$$

with the usual assumption about activity coefficients of the ions. With the introduction of microelectrodes[38] it was possible to confirm this relationship for muscles under the conditions described above[39,40] for $[K]_o$ of 10 mM or greater concentration (Fig. 3.2). At physiological potassium concentrations however the relationship did not hold when the bathing fluid was Ringer; the reason was said to be the net influx of sodium into the

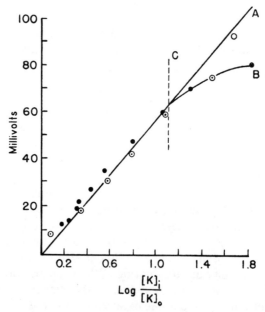

Figure 3.2 Mean membrane potential of frog sartorii plotted against log $[K]_i/[K]_o$. Closed circles give values for muscles immersed in Ringer to which solid KCl had been added for 24 hr at 5°C before E_m was measured at room temperature. Open circles give immediately determined potentials, with acetate replacing chloride and bicarbonate in external fluid. Theoretical line A has a slope of 57 mV and dotted line C represents the level of 10 mM K_0. From E. J. Conway, Physiol. Rev. **37**, 84 (1957).

fibers, which could be prevented either by using frog plasma as immersion fluid or by incorporating some plasma in the Ringer used.[41,42] Under these conditions the above relationship held even at the physiological potassium concentration in the case of both frog[43] and rat muscles.[44] In order to quantify the depolarizing effect of sodium ions under these conditions in terms of relative permeabilities of sodium and potassium ions, the Goldman constant field equation was applied[40-45] on the assumption that steady-state conditions existed. This has been used in the form

$$E_m = - \frac{RT}{F} \ln \frac{P_K[K]_i + P_{Na}[Na]_i + P_{Cl}[Cl]_o}{P_K[K]_o + P_{Na}[Na]_o + P_{Cl}[Cl]_i} \quad (17)$$

When the ratio P_{Na}/P_K was assumed to be 1/100 with a rapid equilibration of chloride ions with the potential,[40,45] there was good agreement between the diffusion potential calculated with this equation and that measured in frog muscle fibers in normal Ringer fluid with 2.5 mM potassium.

Isolated single fibers of frog muscle were used for the measurement of rapid changes in membrane potential in response to similar changes in the composition of the bathing fluid. With these, delay due to diffusion of ions through interstitial fluid was absent; and the rates of potential changes provided information on the kinetics of the ion fluxes, revealing, for example, the relative permeabilities of potassium and chloride. When $[K]_o$ and $[Cl]_o$ were changed reciprocally so that their product was constant or when $[K]_o$ was changed in Cl-free solution, the membrane behaved like a potassium electrode. At constant $[K]_o$, on the other hand, changing $[Cl]_o$ produced only a transient displacement of E_m in the direction expected for a chloride electrode. At constant $[Cl]_o$, increasing $[K]_o$ from 2.5 to 10 mM produced a sudden depolarization from -94 to -73 mV, which was immediately reversible if the fibers were returned to normal Ringer within about 8 sec. When the fibers were exposed to 10 mM K for about 40 min (Fig. 3.3), the E_m drifted to equilibrium at -65 mV. On returning to 2.5 mM K, the immediate repolarization was only 3 mV, and it took about 40 min to reach -90 mV. When the depolarizing fluid contained 95 mM K_o, on returning the fibers to 2.5 mM K no repolarizing took place, but the membrane then behaved like a chloride electrode so that E_m changed to $+64$ mV as $[Cl]_o$ was reduced to 3.6 mM.

These experiments uncovered a phenomenon of anomalous rectification in muscle,[46] that is, a greater permeability for inward than for outward movement of potassium ions. While potassium and chloride ions carry current across the fiber membrane under the conditions described, their relative contribution to the total current depends on the direction in which potassium is moving. Potassium permeability for inward current

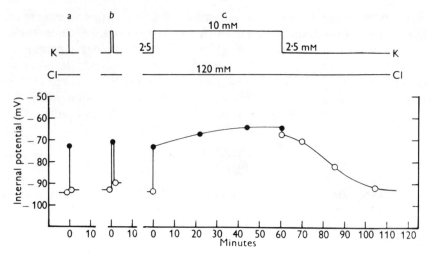

Figure 3.3 Effect on E_m of changing $[K]_0$ from 2.5 to 10 mM at constant $[Cl]_0$. Open circles observed with 2.5 mM K. Results (a) and (b) were obtained on fibers of 119 and 173 μm in diameter, respectively. From A. L. Hodgkin & P. Horowicz, *J. Physiol.* **148**, 127 (1959).

was about 8×10^{-6} cm·sec^{-1} but was only 0.05×10^{-6} cm·sec^{-1} for outward current, compared with 4×10^{-6} cm·sec^{-1} for chloride conductance. In the resting fiber, chloride conductance appeared to be about twice the potassium conductance, their absolute magnitude being 100 and 200 μmho·cm^{-2}, respectively.

It is evident from eq. (17) that the diffusion potential depends not only on the concentrations or activity of diffusible ions inside and outside the cell but also on the relative magnitude of the permeability coefficients P_K, P_{Na}, and P_{Cl}. Where the membrane was permeable to only one of these ionic species to the exclusion of the other, membrane potential was determined by the Nernst equation, eq. (1), with respect to the particular ion. For example in the cells of the isolated guinea-pig taenia coli,[47] the equilibrium potentials were calculated to be -89 mV E_K, $+52$ mV E_{Na}, and -24 mV E_{Cl}. As the "resting potential" (the most negative potential reached) was about -53 mV in this spontaneously active smooth muscle, none of these ions was at equilibrium and all had to be actively pumped to maintain steady-state conditions. The values of the terms P_K, P_{Na}, and P_{Cl} for use in the Goldman equation have been estimated from the efflux of ^{42}K, ^{22}Na, and ^{36}Cl from labeled muscle into inactive physiological saline. Results were plotted as logarithms of the decrease in tissue activity as a function of time, and by extrapolation of the linear part of the graph back to the ordinate an estimate of the amount of labeled potassium and chloride within the cells was obtained. While this procedure seemed to be valid for these ions, the nonlinearity of the graph for sodium raised a problem. The efflux of potassium, m_0^K, like that of the other ions was calculated by the equation

$$m_0^K = kC_i \frac{V}{A} \tag{18}$$

where k is the rate constant for the flux, C_i is the intracellular concentration of the label, and V/A is the mean volume/surface area ratio of the cells, which was about 1.4 μ. The values found in taenia coli were: m_K, 4 pmole·cm^{-2}·sec; m_{Cl}, 8.4 pmole·cm^{-2}·sec; and m_{Na}, 7.2 pmole·cm^{-2}·sec which agreed well with influx rates measured under similar conditions. A correction was applied for the effect of diffusion through the extracellular space on measured fluxes. Ideally, to avoid this problem, measurements are made on isolated single fibers of nerve or muscle, but this was not possible with smooth muscle.

Following the efflux measurements the amount of radiation in the tissue at the end of the efflux period was added to the activity of successive samples of effluent in reverse order, giving the amount of activity in the fibers as a function of time (count·min^{-1}). The radioactivity leaving the muscle in unit time over the washout period and expressed in count·min^{-2} was also determined. Both these were plotted logarithmically against time since their comparison was an indication of the degree of homogeneity of the label within the muscle. The two curves were quite parallel in the case of potassium and chloride but not in the case of sodium (Fig. 3.4). This was due apparently to the large loss of extracellular sodium at the early stages of efflux and perhaps due to a slowly exchanging fraction toward the end of the washout period. To help resolve the curves, after labeling with ^{22}Na the muscle was soaked for 15 min in inactive Ringer at 4°C to allow time for washout of the extracellular space before transmembrane efflux was initiated by raising the temperature of the preparation to 35°C. This helped establish a linearity of the curve, which facilitated extrapolation for estimation of initial intracellular sodium counts.

The permeability constants were then calculated by the constant field assumption

$$P_K = \frac{m_K}{[K]_i \times \frac{EF/RT}{1 - e^{-EF/RT}}} \tag{19}$$

where the term $(EF/RT)/(1 - e^{-EF/RT})$ represents the effect of the electrical field on the movement of the ion and amounted to 0.3 for the outward movement of K and 2.46 for the inward movement of sodium and outward

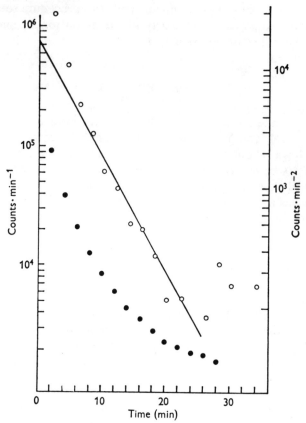

Figure 3.4 ^{22}Na efflux fom taenia coli at 35°C. Open circles show decrease in activity in the effluent and filled circles show the decrease in activity left in tissue. From R. Casteels, *J. Physiol.* **205**, 193 (1969).

movement of chloride at E_m −55 mV and 35°C. The values found were P_K 11 × 10⁻⁸ cm·sec⁻¹, P_{Cl} 6.7 × 10⁻⁸ cm·sec⁻¹, and P_{Na} 1.8 × 10⁻⁸ cm·sec⁻¹.

Sodium permeabilities of skeletal muscles and smooth muscle did not seem to be significantly different, but P_K and P_{Cl} were much lower in the latter. In the former, P_{Cl} was 4 × 10⁻⁶ cm·sec⁻¹ and P_K (where E_m and E_K were close together) was 1.6 × 10⁻⁶ cm·sec⁻¹.

Introducing these values into the Goldman equation gave a diffusion potential of −37 mV for smooth muscle. The difference between the measured potential of −53 mV and the calculated value was attributed to a direct effect of the sodium pump on membrane potential, namely, an electrogenic pump, which will be discussed later. It is necessary to specify the magnitude of the force driving potassium across the membrane and also the direction of the potassium movement under these conditions because of the phenomenon of anomalous rectification.[46] Before dealing with this question it might be worthwhile to consider the relationship between the permeability coefficients and membrane conductance to various ions. Potassium conductance may be expressed as follows:

$$g_K = \frac{P_K F^3 E [K]_0}{R^2 T^2 (1 - e^{-EF/RT})} \text{ mho} \cdot \text{cm}^{-2} \qquad (20)$$

Although the terms permeability and conductance have been used as though they were interchangeable in relation to potassium movement through the cell membrane and although their dimensions are similar (cm·sec^{-1}), the relationship between these two terms can be modified by the distribution of potassium ions across the cell membrane as evidenced by a further relationship, namely,

$$g_K = P_K \frac{F^3 E}{(RT)^2} \frac{[K]_0 [K]_i}{[K]_0 - [K]_i} \qquad (21)$$

Potassium conductance may also be expressed in more general terms for the cases in which E_m and E_K are not identical:

$$g_K = \frac{P_K F^2}{RT} \frac{E_m}{E_K - E_m} \frac{[K]_0 - [K]_i e^{-EF/RT}}{1 - e^{-EF/RT}} \qquad (22)$$

or in terms of membrane fluxes as follows:

$$g_K = \frac{F^2}{RT} \frac{m_K^o - m_K^i e^{-EF/RT}}{1 - e^{EF/RT}} - \frac{F}{E}(m_K^o - m_K^i) \qquad (23)$$

where m_K^o and m_K^i represent potassium efflux and influx, respectively. Its relation to the ionic current density and driving force across the cell membrane is defined as

$$I_K = g_K(E_m - E_K) \qquad (24)$$

where I_K is current density (A·cm^{-2}).

ANOMALOUS RECTIFICATION IN MUSCLE

Returning to the important question of anomalous rectification in skeletal muscle fibers, this was exemplified in their ability to pass a large transmembrane current in high-potassium solution if the potential gradient was inward, that is, E_m more negative than E_K, but very little current when the gradient was outward, or E_K more negative than E_m. For example,[45] in muscle fibers in 95 mM potassium ($E_K = -22$ mV), the conductance for inward potassium current at E_m of -32 mV was 3000 μmho·cm^{-2}, whereas that for outward current in 2.5 mM ($E_K = -93$ mV) and E_m of -64 mV was only 30 μmho·cm^{-2}. The potassium conductance at equilibrium was about 1000 μmho·cm^{-2}.

The rates of net efflux of potassium and chloride ions from frog sartorii which were first loaded with these ions by immersion for about 4 hr in Ringer containing 100 mM KCl and then reimmersed in normal Ringer confirmed[48] the slow exit of potassium suggested by the anomalous rectification of the membrane. After 2 hr in the 2.5 mM K reimmersion fluid, $[Cl]_i$ had decreased from about 100 to 77 m.mole·l^{-1} fiber water, while $[K]_i$ had only fallen from 222 ± 7.3 to 216 ± 4.5 m.mole·l^{-1} fiber water during this period. A voltage clamp was applied[49] through microelectrodes placed near the ends of the fibers within intact isolated sartorii of frog (Fig. 3.5). The initial current on applying the pulse was increased during hyperpolarization and decreased by depolarization of about 20 mV. The changes in conductance were in line with anomalous rectification and were insensitive to the presence or absence of chloride ions in the medium, although the absolute conductance was reduced in the absence of chloride. For small deviations of potential from the resting level, about two thirds of the membrane current was carried by chloride and one third was carried by potassium.

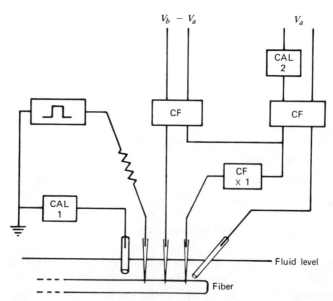

Figure 3.5 Three-microelectrode arrangement for recording membrane potential and membrane current near the end of a single muscle fiber in response to constant current pulses through microelectrode furthest from end of fiber. Microelectrodes were about 445 μm apart. E_m changes at two microelectrodes connected to cathode followers (CF) in response to current pulse were used in calculation of membrane conductance. From R. H. Adrian & W. H. Freygang, *J. Physiol.* **163**, 61 (1962).

To explain the phenomenon of anomalous rectification, the following suggestions were made: (a) the surface of the muscle fiber at rest had a very low K permeability but a large and constant Cl permeability; (b) the greater part of the resting K permeability was in the walls of a part of the sarcotubular system in contact with the external fluid; and (c) the membrane between the lumen of this component and the sarcoplasm was the site of K permeability changes underlying anomalous rectification. The conductances measured were consistent with the presence of two parallel channels for K rectifying in opposite directions and a single Cl channel. There appeared to be a restriction in the movement of potassium but not of chloride, which would be explicable on the basis of an accumulation of potassium within the lumen of the transverse-tube and associated sarcotubular space. The extent of the triad elements in skeletal muscle fibers has been measured by the use of large molecules such as ferritin[50] and horse radish peroxidase.[51] These substances fill the central element of the triad, that is, the transverse tubule, when introduced into the fluid bathing the muscle (see Plate 3).

If restricted diffusion in the T-system is in fact responsible for anomalous rectification, then destruction of this system and inactivation of internal membranes might be expected to change the resting permeability of the fiber and abolish anomalous rectification. This has been done by soaking muscle fibers for about 1 hr in Ringer made hypertonic with glycerol and then reimmersing them in normal Ringer. After this treatment, which also abolished excitation–contraction coupling, it has been found that only 2.6% of the sarcomeres had t-tubes containing peroxidase, compared with 98.5% in normal muscles. At the same time membrane capacitance, which is generally taken to be a function of membrane surface, was reduced.[52] Such drastic treatment led to the depolarization of some of the muscle fibers, but it was found[53] that if the concentration of calcium and magnesium in the reimmersion fluid was raised, fibers maintained normal potentials for up to 6 or 8 hr in spite of the disruption of the internal tubular system. On returning frog muscles to normal Ringer solution after the glycerol treatment,[51] the terminal cisternae and longitudinal tubes near the middle of the sarcomere were almost normal, while the transverse tubule and intermediate cisternae (the reticulum near the junction of the A and I bands were either absent or disintegrated, often appearing as vacuoles (see Plates 4a, b, and c). Ferritin no longer enters the transverse tubule.[54,55]

When glycerol-treated single muscle fibers[55] made potassium rich in 165 mM KCl were reimmersed in 40 mM K, Cl-free Ringer, they repolarized immediately, unlike normal fibers with an intact T-system, so anomalous rectification evidently depended on the integrity of the internal membranes.

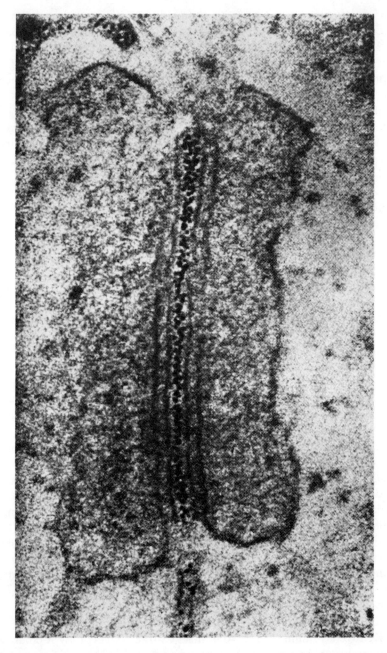

Plate 3 Typical appearances of the triad in ferritin-treated muscles with central element packed full of ferritin molecules. From H. E. Huxley, *Nature* (Lond.) **202,** 1067 (1964).

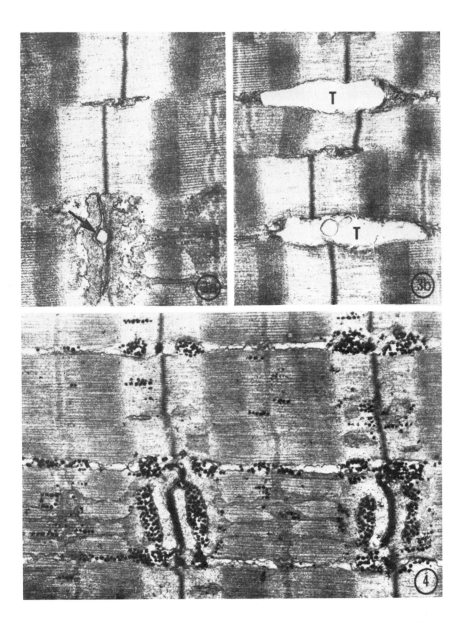

Plate 4 (a) Frog skeletal muscle after incubation for 30 min in 400 mM glycerol-Ringer solution and 30 min in normal Ringer solution, showing localized swelling of a TT (arrow). (b) Disrupted TT under same conditions showing no clear membrane limit with longitudinal system. (c) Frog muscle incubated for 30 min in Ringer, followed by 30 min in 400 mM glycerol-Ringer containing 3 mg·ml^{-1} horseradish peroxidase and then fixed. Note normal structure of the T system and the tracer substance filling all TT. ×30,000. From G. Niemeyer & W. G. Forssmann, *J. Cell Biol.* **50**, 288 (1971).

The study of the role of the T-system in resting potassium conductance has been facilitated by steps taken to reduce the chloride conductance of the sarcolemma. Chloride conductance of frog sartorius muscles fibers[56,57] was decreased markedly in acidic Ringer and increased in alkaline conditions over the pH range of 5 to 9.8. Ionic conductances in resting sartorius muscle fibers of frog were estimated from input resistance measurements[58] under control conditions, where chloride permeability has been reduced by lowering the pH and where the T-system had been disrupted as described, in an attempt to quantify the conductance of potassium and chloride at the surface membrane, g_K^s and g_{Cl}^s, respectively, and in the tubular system g_K^t and g_{Cl}^t, respectively. The values found for these were g_K^t 55 μmho·cm^{-2}, g_K^s 28 μmho·cm^{-2}, g_{Cl}^s 219 μmho·cm^{-2}, and g_{Cl}^t 0 μmho·cm^{-2}, indicating the confinement of chloride conductance to the sarcolemma and of most of the potassium conductance to the transverse tubule and associated structures. However these studies also indicated that the surface membrane might exhibit anomalous rectification.

In addition to the reduced potassium conductance following application of depolarizing electrical current, a similar decline has been observed when hypolarizing current of long duration was used on muscle fibers in normal Ringer. The latter has been attributed on the one hand[59] to a decrease in membrane potassium permeability and on the other[60] to depletion of this ion within the T-system following its uptake into the sarcoplasm from a space suggested to occupy no more than 3% of the fiber volume. When the hyperpolarizing current was switched off, a recovery of conductance took place which could be quantified by applying a hyperpolarizing pulse (test pulse) at various intervals after the first (conditioning) hyperpolarizing pulse had been switched off. It was found[60] that recovery to normal conductance at room temperature could be resolved into two rates that were dependent on the magnitude of the initial hyperpolarization. For example, when the conditioning current raised E_m above -120 mV, the rate of recovery was fast (<200 msec) and had the characteristics of a permeability change such as a high temperature coefficient (about 5.6). For lesser hyperpolarizations the recovery rate was slow ($t_{1/2}$ about 400 msec) and Q_{10} was about 1.5, similar to that of a diffusion process. When prolonged hyperpolarization to potentials less negative than -120 mV was carried out and then removed, the resulting ionic current at the resting potential was outward and lasted for several seconds and E_m at equilibrium had become more negative by 3–5 mV (Fig. 3.6). These observations suggested that a decline of tubular K concentration was responsible for the fall in conductance during hyperpolarizing potentials less than -120 mV and that the replenishment of this potassium by diffusion was responsible for recovery of membrane current. It was esti-

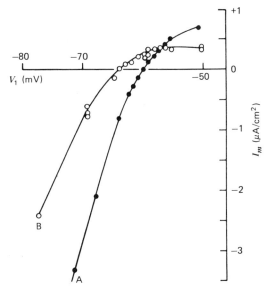

Figure 3.6 Instantaneous current–voltage curves from the resting potential (A) and following 2.5 sec hyperpolarization to $E_2 = -110$ mV (B). Currents were measured 50 msec after potential changes. From W. Almers, *J. Physiol.* **225**, 33 (1972).

mated[61] that the inward K current during hyperpolarization reduced its concentration in the tubules by 50% and that the space containing this potassium was less than 0.8% of the fiber volume.

The permeability changes believed to occur following larger hyperpolarizations appeared[62] to be sensitive to the replacement of most of the sodium in the bathing fluid with tetramethylammonium (TMA) sulfate. It has therefore been suggested that sodium driven into the membrane during the hyperpolarization might react with binding sites, thereby blocking potassium conductance.

BLOCKING OF POTASSIUM CHANNELS

The permeability of muscle membranes to potassium ions has also been modified by a number of ions and compounds which may interact with the potassium channels. The foreign cations caesium and rubidium slow the rate of movement of potassium through the membrane.[63,64] The action of caesium has been studied by comparing its effect on ^{42}K efflux with that of potassium itself. The rate constant for ^{42}K efflux into potassium-free frog Ringer was found[64] to be 0.061 ± 0.002, increasing to 0.143 ± 0.006 in the presence of 2.5 mM K_0 and decreasing to 0.046 ± 0.003 in the presence of 2.5 mM Cs_0. Under these conditions the depolarizing effects of the two ions were similar in the presence or absence of sodium pump inhibition by ouabain. The depolarizing action of Cs was not due to inward movement of this cation, which in the presence of ouabain is only one hundredth that of potassium at the same concentration. In fact the P_{Na}/P_K ratio appeared to increase threefold in the presence of Cs mainly owing to a fall in P_K.

In voltage-clamped squid axons,[66] Cs^+, Na^+, Li^+, and to a lesser degree Rb^+ inhibited the exit of potassium. Tetraethylammonium ion (TEA) also had a similar effect when applied to the inside of the membrane but not when applied outside. The inward facing part of the potassium channel seems to have a wider mouth (8 Å) than the outer one (3 Å) and therefore can accept TEA and similar large ions which however cannot pass to the external fluid.[67] The nonyltriethylammonium derivative appeared to be more effective than TEA itself in blocking the K channels of squid axon. It was suggested[68] that the similarity in size of the ethylated part of the molecule to the hydrated potassium ion coupled with the hydrophobic nature of the C_9 hydrocarbon chain were responsible for its effectiveness when applied on the inside of the axon. A depolarizing current applied by voltage clamp seemed to open a gate on the inner wide mouth of the pore, allowing K ions to enter the pore accompanied by the C_9 compound which eventually inactivated the pore, causing decline of outward K current. With return of membrane potential to its resting value the gates probably closed once more, trapping the C_9 compound within the membrane. Recovery of K conductance may then be achieved either by increasing $[K]_0$ or by making E_m more negative, thereby driving K ions into the pores from the external solution. When $[K]_0$ was 440 mM, the half-time for recovery was 6 msec, compared with 36 msec in 10 mM K. The entry of potassium into the pore from outside ejected the C_9 compound by electrostatic repulsion.

Both the sodium and lithium ions appeared to be effective in reducing K conductance when applied inside the membrane. In the case of sodium-enriched muscle fibers, the fall in potassium permeability was evident in that the membrane potential changed by only 25 mV for a 10-fold change in $[K]_0$ but by 48 mV for a similar change in $[Cl]_0$.[69] In voltage-clamped axons[68] addition of 100 mM sodium to the internal perfusion medium inhibited the outflow of K^+ through the K channels (Fig. 3.7).

When added to frog skeletal muscle immersed in normal Ringer, barium[70] caused a slight rise in membrane resistance. In Cl-free Ringer with sucrose substitution, R_m increased from 9.7 ± 0.6 kΩ·cm^{-2} to 29.5 ± 3.0 kΩ·cm^{-2} after addition of 0.5 mM Ba^{2+} and increased significantly even in the presence of as little as 0.05 mM Ba, indicating a fall in K conductance.

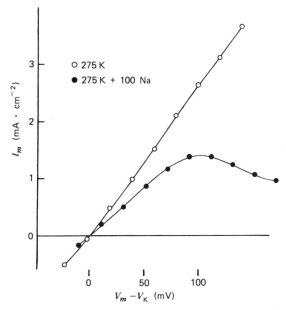

Figure 3.7 Addition of sodium to the internal perfusion medium inhibits efflux of K⁺ through K pores. Curves are instantaneous I–V curves in TTX-treated axons, with g_K fully activated by depolarizing prepulse. From C. M. Armstrong, Q. Rev. Biophys. **7**, 179 (1974).

A three-electrode voltage clamp was used[71] to examine the effects of barium and strontium on inwardly rectifying K conductance of resting frog sartorius fibers. When 0.01–5 mM Ba^{2+} was added to the control solution (115 mM K), the inward current during a hyperpolarizing step decreased exponentially with time as Ba^{2+} blockade developed. The barium ion decreased the steady-state and instantaneous inward current recorded during hyperpolarization. The degree of blockade was dependent on membrane potential, steady-state current being increasingly inhibited with increasing hyperpolarization. The concentration of barium needed for 50% blockade decreased therefore with increasing hyperpolarization and the apparent dissociation constant showed an e-fold decrease for a 16.8 mV increase in membrane potential.

Barium ions were about 400 times more effective than Sr ions in blocking the inward potassium current; but while inhibition of this current by external Cs was instantaneous, that by Ba developed more slowly following hyperpolarization of the membrane. Potential-dependent blockade by ions and polar substances are not uncommon, having been described for TEA and aminopyridine in the K channels[72] of squid axons and for Cs[73] in the inward rectifying K current of skeletal muscles. The potential dependence of the blockade has been attributed to a gating mechanism of conductance in the channels which prevents access of the blocking agent to binding sites therein. A high temperature coefficient found for blockade by Ba^{2+} suggests a reaction between the barium ion and receptor sites in the channel but not necessarily with carrier sites for potassium. There seemed to be competition between barium and potassium, the barium having access to the site from the outside only but not being capable of passing through the channel.

FACTORS INCREASING POTASSIUM CONDUCTANCE

A large increase in resting potassium conductance has been observed in frog sartorious muscle fibers during metabolic exhaustion,[74] which has been attributed to change in the gating of activated potassium channels responsible for delayed rectification to a permanently open state. The exhaustion was induced by addition of cyanide and iodoacetate to the fluid bathing the muscles and by electrical stimulation to the point where twitch tension had declined to zero. The membrane potential fell from -75 mV to -69 mV and then rose to -83 mV, possibly because of increases in the P_K/P_{Na} ratio, while membrane resistance fell to 4–6% of its original value (to 58 $\Omega \cdot cm^{-2}$). The ratio g_K/g_{Cl} changed from 2:3 in normal fibers to 5:1 in exhausted ones, while in absolute terms g_K appeared to increase from 130 to 14,300 $\mu mho \cdot cm^{-2}$ and g_{Cl} from 200 $\mu mho \cdot cm^{-2}$ to 2900 $\mu mho \cdot cm^{-2}$.

In red blood cells, membrane potential was close to the equilibrium potential for chloride ions, that is, about -14 mV.[75] This is not surprising in view of the high permeability of their membranes to chloride. The equilibrium potential for potassium however was probably similar to that of nerve and some muscles, namely, about -70 to -80 mV, so the red cell membrane was relatively impermeable to potassium. When the potassium ionophore valinomycin (10^{-5} M) was added[76] to the suspending fluid, the resting potential of erythrocytes increased significantly owing apparently to increased permeability and exit of potassium ions from the cells, driven by the potential difference $E_m - E_K$. Likewise when red cells were penetrated more than once, they were also found to become hyperpolarized, and perhaps in this case the increased K permeability was induced by entrance of calcium ions into the cells.

Increased K conductance in red cells following calcium entry has been indicated by the following experiments.[76] When the divalent ionophore A23187 (0.1–0.3 μM) was added to human erythrocytes suspended in solution containing 1 mM ^{45}Ca, the uptake of the ^{45}Ca doubled while $[K]_i$

decreased from about 100 mM to about 70 mM and there was a decrease in cell volume as judged by decrease in osmotic fragility. In the presence of 3.0 to 10 μM ionophore, ^{45}Ca uptake increased 30 to 50 times in toad erythrocytes, and there was a loss of intracellular potassium which was prevented by increasing $[K]_0$ to 90 mM.

Similarly in the large neurons of the snail *Helix aspersa*[77] there was increased potassium permeability in response to increase in intracellular calcium concentration. When the latter was injected into the neuron, there was a fall in membrane resistance and the membrane became hyperpolarized. The reversal potential for this response was dependent on the external concentration of potassium. A similar response was seen[78] in motoneurones of the cat, where the increase in K conductance led to a decreased excitability.

The calcium ionophore A23187 has also been used in the lacrimal gland[79] of rat to demonstrate increased rate of potassium release (^{86}Rb). It had been found that adrenaline brought about a large increase in the rate of Rb efflux, which was sustained in the presence of calcium outside but was only a transient event in its absence.

REFERENCES

1. J. V. Evans & A. T. Phillipsen, *J. Physiol.* **139**, 87 (1957).
2. G. N. Ling, *A Physical Theory of the Living State: The Association–Induction Hypothesis*, Blaisedell Publ. Co., Waltham, Mass., 1962.
3. G. N. Ling, *Mol. Cell Biochem.* **15**, 159 (1977).
4. H. J. Curtis & K. S. Cole, *J. Gen. Physiol.* **27**, 649 (1939).
5. G. Marmont, *J. Cell Comp. Physiol.* **34**, 351 (1949).
6. A. L. Hodgkin & A. F. Huxley, *J. Physiol.* **117**, 500 (1952).
7. E. J. Conway, *Physiol. Rev.* **37**, 84 (1957).
8. A. L. Hodgkin, The Croonian Lecture, *Proc. Roy. Soc.* **B148**, 1 (1958).
9. G. Gardos, *Acta Physiol. Hung.* **6**, 191 (1954).
10. R. D. Keynes & R. A. Steinhardt, *J. Physiol.* **198**, 581 (1968).
11. H. H. Ussing, *Physiol. Rev.* **29**, 127 (1949).
12. A. Chivet, T. Clausen & L. Giradier, *J. Physiol.* **266**, 43 (1977).
13. E. J. Conway, R. P. Kernan & J. A. Zadunaisky, *J. Physiol.* **155**, 268 (1961).
14. L. Minkoff & R. Damadian, *Biophys. J.* **13**, 167 (1973).
15. S. H. White & K. L. Ibsen, *Biophys. J.* **13**, 1001 (1973).
16. J. A. Raven, *Biophys. J.* **13**, 1002 (1973).
17. A. Pena, *Arch. Biochem. Biophys.* **167**, 397 (1975).
18. M. Cockburn, P. Earnshaw & A. A. Eddy, *Biochem. J.* **146**, 705 (1975).

19. R. Damadian, *Science* **165**, 79 (1969).
20. J. P. Caille & J. A. M. Hinke, *Can. J. Physiol. Pharmacol.* **51**, 390 (1973).
21. J. H. Wang, *J. Am. Chem. Soc.* **76**, 4755 (1954).
22. J. P. Chaille & J. A. M. Hinke, *Can. J. Physiol.* **50**, 228 (1972).
23. M. J. Kushmerick & R. P. Podolsky, *Science* **166**, 1297 (1969).
24. G. N. Ling, *J. Physiol.* **280**, 105 (1978).
25. E. Rogus & K. L. Zierler, *J. Physiol.* **233**, 227 (1973).
26. R. L. Birks & D. F. Davey, *J. Physiol.* **202**, 171 (1969).
27. R. P. Kernan & M. MacDermott, *J. Physiol.* **263**, 158P (1976).
28. G. Ling & M. H. Kromash, *J. Gen. Physiol.* **50**, 677 (1967).
29. D. A. T. Dick & S. G. A. McLaughlin, *J. Physiol.* **205**, 61 (1969).
30. J. A. M. Hinke, *J. Physiol.* **156**, 314 (1961).
31. A. A. Lev, *Nature* (Lond.) **201**, 1132 (1964).
32. L. G. Palmer, T. J. Century & M. M. Civan, *J. Membr. Biol.* **40**, 25 (1978).
33. L. N. Vorobiev, *Nature* (Lond.) **216**, 1325 (1967).
34. P. J. Boyle & E. J. Conway, *J. Physiol.* **100**, 1 (1941).
35. L. J. Mullins, *J. Gen. Physiol.* **42**, 817 (1959).
36. O. F. Hutter & A. E. Warner, *J. Physiol.* **189**, 427 (1967).
37. W. O. Fenn, *Physiol. Rev.* **16**, 442 (1936).
38. J. Graham & R. W. Gerard, *J. Cell. Comp. Physiol.* **29**, 99 (1946).
39. R. P. Kernan & E. J. Conway, Abstr. 3rd Inter. Biochem. Congr. 9, 29, 1955.
40. R. H. Adrian, *J. Physiol.* **133**, 63 (1956).
41. M. J. Carey & E. J. Conway, *J. Physiol.* **125**, 232 (1954).
42. R. Creese, *J. Physiol.* **197**, 255 (1968).
43. R. P. Kernan, *Nature* (Lond.) **185**, 471 (1960).
44. R. P. Kernan, *Nature* (Lond.) **200**, 474 (1963).
45. A. L. Hodgkin & P. Horowicz, *J. Physiol.* **148**, 127 (1959).
46. B. Katz, *J. Physiol.* **163**, 61 (1949).
47. R. Casteels, *J. Physiol.* **205**, 193 (1969).
48. R. H. Adrian, *J. Physiol.* **151**, 154 (1960).
49. R. H. Adrian & W. H. Freygang, *J. Physiol.* **163**, 61 (1962).
50. H. E. Huxley, *Nature* (Lond.) **202**, 1067 (1964).
51. B. Eisenberg & R. S. Eisenberg, *Science* **160**, 1243 (1968).
52. R. S. Eisenberg & P. W. Gage, *Science* **158**, 1700 (1967).
53. R. S. Eisenberg, J. N. Howell & P. C. Vaughan, *J. Physiol.* **215**, 95 (1971).
54. S. A. Krolenko, *Nature* (Lond.) **221**, 966 (1969).
55. S. Nakajima, Y. Nakajima & L. D. Peachey, *J. Physiol.* **200**, 115 (1969).
56. O. F. Hutter & A. E. Warner *J. Physiol.* **189**, 403 (1967).
57. N. C. Spurway, *J. Physiol.* **181**, 51P (1965).

58. R. S. Eisenberg & P. W. Gage, *J. Gen. Physiol.* **53,** 279 (1969).
59. R. H. Adrian, W. K. Chandler & A. L. Hodgkin, *J. Physiol.* **208,** 607 (1970).
60. W. Almers, *J. Physiol.* **225,** 33 (1972).
61. W. Almers, *J. Physiol.* **225,** 57 (1972).
62. N. B. Standen & P. R. Stanfield, *J. Physiol.* **282,** 18P (1978).
63. R. A. Sjödin, *J. Gen. Physiol.* **42,** 983 (1959).
64. V. Bolingbroke, E. J. Harris & R. A. Sjodin, *J. Physiol.* **157,** 289 (1961).
65. L. A. Beaugé, A. Medici & R. A. Sjodin, *J. Physiol.* **228,** 1 (1973).
66. F. Bezanilla & C. M. Armstrong, *J. Gen. Physiol.* **60,** 588 (1972).
67. C. M. Armstrong & B. Hille, *J. Gen. Physiol.* **59,** 388 (1972).
68. C. M. Armstrong, *Q. Rev. Biophys.* **7,** 179 (1975).
69. R. P. Kernan, *J. Gen. Physiol.* **51,** 2045 (1968).
70. N. Sperelakis, M. F. Schneider & E. J. Harris, *J. Gen. Physiol.* **50,** 1565 (1967).
71. N. B. Standen & P. R. Stanfield, *J. Physiol.* **280,** 169 (1978).
72. C. M. Armstrong, *J. Gen. Physiol.* **58,** 413 (1971).
73. J. Z. Yeh, G. S. Oxford, C. H. Wu & T. Narahashi, *J. Gen. Physiol.* **68,** 519 (1976).
74. L. A. Gay & P. R. Stanfield, *Nature* (Lond.) **267,** 169 (1977).
75. R. Fink & H. C. Lüttgau, *J. Physiol.* **263,** 215 (1976).
76. U. V. Lassen & O. Sten-Knudsen, *J. Physiol.* **195,** 681 (1968).
77. R. W. Meech, *J. Physiol.* **237,** 259 (1974).
78. K. Kryjevic & A. Lisiewicz, *J. Physiol.* **225,** 363 (1972).
79. R. J. Parod & J. W. Putney, *J. Physiol.* **281,** 371 (1978).

4
Potassium Fluxes and the Action Potential

Since the classical investigations of Hodgkin, Huxley, and Katz[1,2] on the giant nerve axon of squid, establishing the role of sodium and potassium ion fluxes in the generation of the action potential, the application of the voltage clamp technique to skeletal and cardiac muscles has revealed subtle differences in electrogenesis between the latter and nerve, particularly in relation to potassium currents. In the earlier studies on nerve, impulse propagation was prevented and the direction and magnitude of ionic currents measured across well-defined areas of axon membrane, when the membrane potential was suddenly changed and held at a predetermined value by the electronic feedback circuit. When the potential was increased or slightly decreased in a steplike manner from its normal value, the membrane behaved like a leaky condenser. When the membrane was made more negative inside, an inward ionic current carried mainly by potassium was set up, which, but for the feedback voltage stabilization, would have reduced the potential to its original value.

When on the other hand the membrane was depolarized by about 15 mV by the voltage clamp, the current was outward. When the depolarizing step was 20 mV or more, ionic current flowed inward for about 0.5 msec and was followed by an outward current lasting for as long as the voltage clamp was maintained. Beyond a certain critical depolarization the membrane no longer behaved passively, but the initial response was a depolarizing current, tending to enhance rather than resist the effect of applied current, and the pattern of current change was similar to that expected during the action potential (Fig. 4.1A). When sodium was omitted from the bathing fluid, the inward current was absent, indicating that it was carried by this cation, and the remaining outward potassium current was about $1\ mA \cdot cm^{-2}$ (Fig. 4.1B). By subtracting these curves, the magnitude and time course of the sodium current was resolved (Fig. 4.1C). During a single action potential in squid axons, about $3 \times 10^{-12}\ mole \cdot cm^{-2}$ sodium and potassium crossed the cell membrane down their respective electrochemical gradients. This was equivalent to 20,000 $ions \cdot \mu m^{-2}$ cell surface.

HODGKIN–HUXLEY MODEL OF CONDUCTANCE CONTROL DURING THE ACTION POTENTIAL IN NERVE

The Hodgkin–Huxley model of voltage- and time-dependent changes in ionic current flow was put forward to account for the pattern of potential change occurring during the nerve action potential. This theoretical model may be summarized as follows. As the membrane potential was made less negative, the fraction of open channels in the membrane increased, this being controlled by a voltage-dependent gating mechanism. The response

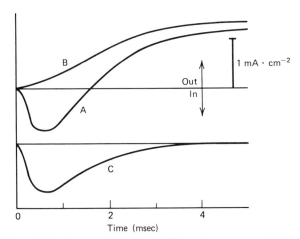

Figure 4.1 Curve A, Membrane currents recorded by voltage-clamp technique following depolarization of the membrane by about 56 mV. Curve B, Outward potassium current recorded when Na_0 was decreased making $E_{Na} = E_m$. Curve C, Inward Na current obtained by subtracting curve B from curve A. From A. L. Hodgkin & A. F. Huxley, *J. Physiol.* **116,** 449 (1952).

of the gates to an instantaneous depolarization was not immediate but required a certain time for the mechanism in the membrane to change to a new steady state. This introduced an element of time dependence into the process which was probably dependent on chemical changes within the membrane, the kinetics of which could be studied.

According to the Hodgkin–Huxley model, each channel was controlled by a charged "gate" which was in either the open α position, allowing the channel to conduct, or the closed β position, which blocks ionic movement. A first-order reaction was assumed between these states. If the fraction of gates in the α state was n, then $1 - n$ represented the fraction in the β state. If the rate coefficient for opening of the gates was α_n and the closing rate coefficient β_n, the rate of opening would be given by $\alpha_n(1 - n)$ and the rate of closing by $\beta_n n$. The net rate of change in the fraction of open channels would then be given by

$$\frac{dn}{dt} = \alpha_n(1 - n) - \beta_n n \tag{25}$$

To account for changes in potassium conductance it was assumed that a channel for this cation was opened when four charged particles had moved to a particular region of the membrane under the influence of the electric field. If n was the probability of a single particle being in the right place, the potassium conductance could be described by the equation $g_K = \bar{g}_K n^4$, where $\bar{g}_K$ was the maximum potassium conductance. It was assumed that three simultaneous events, each with a probability m, opened the channel to sodium and a single event with a probability $1 - h$ closed it. The probability of there being three activating particles and no blocking particles was therefore $m^3 h$. Hence the sodium conductance $g_{Na} = \bar{g}_{Na} m^3 h$, where $\bar{g}_{Na}$ is the maximum sodium conductance. The values of m and h were given by the relationship

$$\frac{dm}{dt} = \alpha_m(1 - m) - \beta_m m \tag{26}$$

and

$$\frac{dh}{dt} = \alpha_h(1 - h) - \beta_h h \tag{27}$$

The α and β term in these equations depend on temperature, calcium concentration, and membrane potential. The effect of depolarization on the membrane was to increase α_m and β_h and to decrease β_m and α_h. The voltage dependence of the gates was due to their charge, which could be influenced by the electric field across the membrane. In the steady state, where $dn/dt = 0$,

$$n_\infty = \frac{\alpha_n}{\alpha_n + \beta_n} \qquad (28)$$

These equations were applied to the voltage clamp data and led to the exponential expressions for n, m, and h and to conductances calculated from the relationships $g_K = \bar{g}_K n^4$ and $g_{Na} = \bar{g}_{Na} m^3 h$. Although a perfect fit was not obtained for the data where the values of α and β were assumed to be voltage dependent, the complete expression for the current density of the membrane was taken to be

$$I = c\frac{dE}{dt} + (E - E_K)\bar{g}_K n^4 + (E - E_{Na})\bar{g}_{Na} m^3 h + (E - E_L)g_L \qquad (29)$$

where the first term represents capacitance current and the last term the current carried by ions other than sodium and potassium. The theoretical action potential obtained by numerical solutions of the various equations for n, m, and h were then plotted (Fig. 4.2).

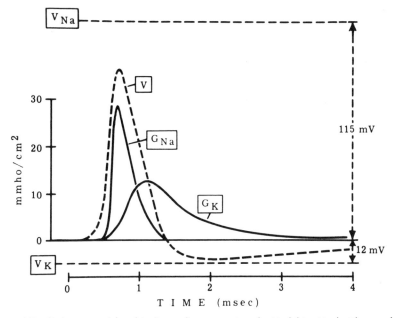

Figure 4.2 Action potential and ionic conductances given by Hodgkin–Huxley theory of the nerve impulse. From A. L. Hodgkin & A. F. Huxley, *J. Physiol.* **117**, 500 (1952).

THE ACTION POTENTIAL OF SKELETAL MUSCLE

In smooth and cardiac muscles, calcium ions contribute to the depolarizing currents[3,4] while at the same time activating the contractile mechanism. The repolarizing currents carried mainly by the potassium ions are more complicated than in nerve. The kinetics of gating mechanism controlling the activation and inactivation of the sodium and potassium channels in the membrane have been extensively studied with a view to understanding the underlying chemical changes within the membrane.

Membrane currents were measured[5,6] by the use of three microelectrodes inserted near the end of the fibers of frog sartorious muscle (Fig. 3.5). One microelectrode was used to deliver current, and the other two were used to measure membrane potential and ionic current. Because contraction of the muscle dislodged the electrodes, excitation–contraction was uncoupled by soaking the muscle in Ringer solution made hypertonic by the addition of sucrose. On applying an appropriate depolarizing step, an early transient inward current abolished by tetrodotoxin and reversing in direction at potentials more positive than +20 mV, was seen. This reverse potential is taken to be approximately equal to the equilibrium potential of the sodium ion. The initial current was followed by a delayed outward current with linear current–voltage relations and an apparent equilibrium potential of −85 mV in normal Ringer. This was called delayed rectifying current and was in the normal direction for repolarization, developing after a delay and inactivating with time.

The variation of the equilibrium potential of the delayed current with respect to changes in [K]$_0$ suggested that the selectivity of the channel for potassium with respect to sodium or P_K/P_{Na} was 30:1, which was considerably less than the value of 100:1 assumed for the resting membrane. Although the role of chloride fluxes in the repolarization process under normal conditions cannot be ignored in skeletal[8] and in cardiac muscle,[7] the use of chloride-free sulfate Ringer in parallel with chloride-containing Ringer has served as some check on their involvement in the process. In normal Ringer, chloride permeability remains high, and this contributes to the electrical stability of the membrane. It seems likely in the condition of myotonia congenita seen in man and in goats[8] that it is reduced surface chloride conductance and to some extent the accumulation of potassium in the transverse tubules[10,11] which brings about a repetitive firing of action potentials and afterdepolarization with consequent muscle tone. A similar response has been seen in goat muscle in Cl-free solution following constant inward current.

The voltage clamp technique was much more satisfactory for studying the late potassium currents than the large regenerative inward sodium current. The latter was eliminated and the delayed currents studied in isolation by treating the muscles with 10^{-6} g·ml^{-1} tetrodotoxin, which blocks the sodium channels. The temperature of this medium was reduced to 3°C to help resolve the currents. The variation of the potassium reversal potential with the amount of current that passed through the delayed current channel was examined[6] with a view to determining the relative contributions of surface and tubular current to the delayed rectification. In squid axons, for example, it had been found[12] that the apparent equilibrium potential for potassium became less negative following depolarizing pulses of long duration, owing apparently to potassium leaving the axon and accumulating in a region of restricted diffusion in a space between the axolemma and the Schwann cell. In the case of the muscle fiber, a similar effect might be expected if most of the potassium current during delayed rectification was into the transverse tubules, which have been estimated to have a volume only about 0.3% of the fiber volume[13] (see Plate 5). After pulses of long duration (400–500 msec) it was in fact found that E_K became less negative by about 10 mV, corresponding to an increase in $[K]_0$ of 4.5 mM. In relation to the calculated efflux of potassium through repolarizing channels, this would represent a volume of restricted diffusion of one sixth to one third of the fiber volume. As this was excessive, it was concluded that most of the delayed current passed through the surface membrane.

Unlike in squid axon, the delayed current in frog muscle inactivated fairly rapidly and almost completely[14] during prolonged depolarization, following an exponential course with a time constant of 0.6 sec at −24 mV or 1 sec at −40 mV (20°). Steady-state inactivation of the potassium current was similar to that of the sodium current but with a less steep voltage dependence. There was considerable variation in the magnitude of the delayed currents and in their rates of development, the cause of which was unknown but may have been due to prolonged immersion in hypertonic solutions.

It was suggested that the lower selectivity of the delayed rectifying channels for potassium to sodium explained the characteristic afterpotential seen in muscle. This appears as a slow phase of repolarization (Fig. 4.3), which sets in when this process is about 80% complete, and it lasts up to 50 msec at room temperature. Temperature has a marked effect on its duration and amplitude, and below 5°C it disappears. At 35°C it leads to repetitive discharge from the membrane in frog. It was concluded that the potential at the peak of the afterpotential was close to the equilibrium potential of the delayed rectifier in both muscle and squid axons, but in

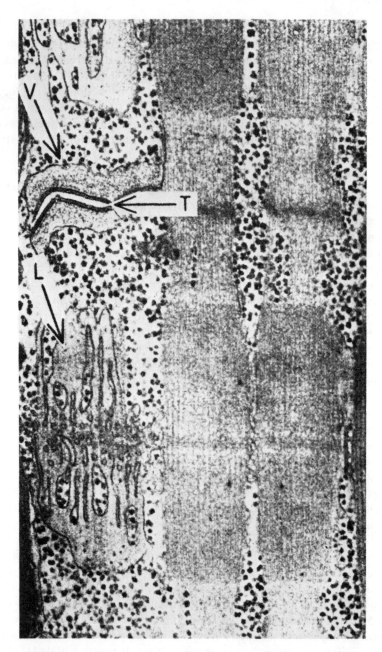

Plate 5 Section of normal frog sartorius muscle passing between two fibrils on the left-hand side of the picture and showing a grazing face-on section of a longitudinal sheet (L) of SR at the level of the A band. On either side of the central channel of the triad (T), the terminal regions of the vesicles (V) of the longitudinal system may be seen. The terminal regions are continuous with the sheet of SR in the center of the sarcomere, but the channel of continuity lies outside the plane of the section. ×37,500. From H. E. Huxley, *Nature* (Lond.) **202**, 1067 (1964).

the latter the resting K selectivity was less than that of the delayed rectifier. Hence the membrane tended to hyperpolarize before return to its steady-state value.

The return of the membrane potential to its resting value depended to some extent on the decline of potassium conductance to its resting level. When ionic currents had been calculated on the basis of the Hodgkin–Huxley model and theoretical action potentials and conduction velocities were calculated for various conditions, assuming that the T-system behaved as a series resistance and capacitance in parallel with the surface capacitance and leak currents, some discrepancies were found between the theoretical and experimental action potentials. The conduction velocity and maximum rate of rise of the theoretical potential were less than those found experimentally. The observed action potential recorded at 2°C had a flatter maximum than the calculated one. Finally the falling phase of the theoretical action potential merged quite smoothly with the negative afterpotential and did not give the dip so often seen in muscle records (Fig. 4.3).

Voltage clamp experiments carried out at 3°C indicated[15,16] that the K current could be subdivided into three components:

1. A fast outward current in the delayed outward rectifying channel which reached its maximum in about 100 msec at -30 mV and declined with a time constant of about 4 msec when the fiber was repolarized to -100 mV. This current had a linear instantaneous current–voltage relation and an equilibrium potential of about -75 mV which was 10–15 mV more positive than the resting potential.
2. A slow current through delayed rectifying channels was detected which reached a peak in 3 sec at -30 mV and declined with a time constant of 500 msec when E_m was returned to -100 mV. This also showed linear instantaneous current–voltage relationship and a mean equilibrium potential of about -83 mV.

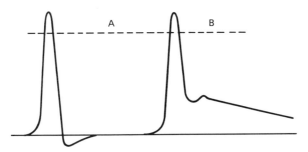

Figure 4.3 Comparison of action potential from nerve axon (A) and skeletal muscle (B) showing afterpotential in the latter.

3. Current was also present in the inward rectifying (anomalous) channel, decreasing with a time constant of 250 msec when the fiber was hyperpolarized to −150 mV. Its equilibrium potential was close to −90 mV, and its current–voltage relationship was typical of inward rectification.

Delayed outward currents which were probably carried by potassium ions were seen also in mammalian muscle under voltage clamp in a double sucrose gap.[17] In many respects these currents, measured in "fast type" fibers of rat skeletal muscle, were similar to those of frog. Two distinct rate constants and equilibrium potentials (reversal potentials) were seen in the outward currents (Fig. 4.4). The "fast" component had a linear instantaneous current–voltage relation, but unlike frog its equilibrium potential was about 6.8 mV more negative than the resting potential. The slow component, which also showed a linear instantaneous current–voltage relation, had an equilibrium potential which was about 26 mV less negative than the resting potential.

The variation in equilibrium potentials has been ascribed on the one hand to the accumulation or depletion of external potassium in a region of restricted diffusion such as the transverse tubules and on the other hand to variations in selectivity of the potassium channels. It is tempting to attribute the more complex pattern of potassium current found in skeletal muscle to its internal membrane system. Although this view is questionable on present evidence, there can be little doubt that a major source of error[16] in the application of voltage clamp to muscles could be the large capacitance currents which appear to flow into the tubular elements through the access resistance during the period of after potential. It has been suggested that this is the current which keeps the surface potential positive with respect to the potassium equilibrium potential. The tail currents seen after a depolarization are made up of tubular capacitance currents and delayed rectifier currents which may be confined mainly to the fiber surface.

GATING CURRENTS FOR POTASSIUM CHANNELS

In this complex of capacitance and ionic currents it would seem to be rather optimistic to seek electrical evidence for the operation of the gating mechanisms, but this has been attempted. It was believed that gating initiated by a change in the electrical field across the membrane might be mediated by a change in electrostrictive pressure across the membrane. However the permeability changes occur within rather narrow limits of membrane potential change, that is, from −70 to −20 mV, whereas

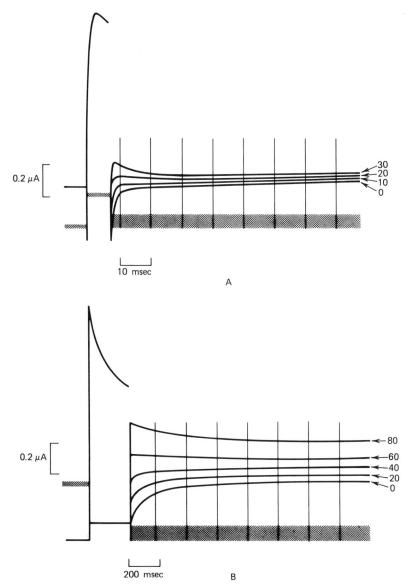

Figure 4.4 Membrane currents and potentials used in measuring reversal potentials of the fast (A) and slow components (B) in rat skeletal muscle at 21°C. From A. Duval & C. Leoty, *J. Physiol.* **278,** 403 (1978).

outside these limits and to the extreme of +200 mV no changes are seen, so that this theory seems less plausible. It has also been suggested that a conformation change in a membrane constituent may be responsible for the gating of sodium and potassium channels. Whatever the trigger for the conformational change, it was suggested that the latter might involve the rotation or movement of molecular dipoles within the membrane structure and it was believed that in the absence of ionic currents such molecular changes might be detected as small current changes distinguishable from capacitance currents by their asymmetry.

The so-called gating currents were investigated in squid giant axons[18] immersed in Na-free solutions in the presence of tetrodotoxin to block inward sodium currents and perfused internally with caesium fluoride to block outward K currents. A signal-averaging technique was also used to filter out the gating currents from the early capacitance currents. Following membrane depolarization by a voltage clamp, an outward current was seen, rising to a peak and declining exponentially to zero. This was followed by an inward current with a similar time course. In addition to the transient currents here, there was an outward current flow during depolarization which had the characteristics of an ionic current with rectification. The magnitude of the charge transfer taking place was unaffected by temperature change, but its time course had a high temperature coefficient, suggesting an intramolecular change. The asymmetric displacement currents here had other properties which suggested that they might represent dipole movements in the membrane; however, it was impossible to relate such charge displacement to the activation of the K channels by the n particles.

Similar measurement made in skeletal muscles[19] suggested a link between these displacement currents and the process of excitation–contraction coupling in that both the current and contraction were increased over the same range of membrane potential with a maximum around −20 mV. The total amounts of charge involved in the "on" and "off" responses to a depolarizing pulse were identical and remained so in various experimental conditions, suggesting that this movement was fixed as expected in the rotation of dipoles. The detection of charge movement was perhaps best accomplished by subtracting the membrane current required for two identically shaped steps, one a control outside the range where gating occurred and the other within the range where charge displacement took place.

Potassium conductance turns on with high-order kinetics and turns off with first-order kinetics.[15] Agreement was therefore sought[20] between charge movement in frog sartorii bathed in a medium designed to minimize all ionic current and this kinetic pattern. One of the uncertainties associated with the use of media containing TTX, TEA, and Rb_2SO_4 is the degree to which these can block conductances in pores without interfering with the gating mechanisms, and another is the extent to which movement of these substances into and out of pores contributes to the observed displacement currents. While some suggestion of K gating current in frog sartorius has been obtained, other experiments[21] have been interpreted as indicating that displacement currents are unrelated to the opening and closing of the K channels. For example, tetracaine and TEA, which slowed the kinetics of the delayed K channels four- to fivefold or shifted the voltage dependence of the delayed channels by 25 mV in a positive direction, had little or no effect on displacement current.

When the three-microelectrode voltage clamp technique was applied[22] to fibers of frog sartorius muscle bathed in hypertonic solutions containing Rb_2SO_4, $CaSO_4$, $(TEA)_2SO_4$ to suppress delayed rectifier current, and 10^{-2} M tetrodotoxin to suppress sodium current, the whole being buffered by phosphate at pH 7.2, it was assumed that ionic currents were absent and that whatever electrical activity remained was a manifestation of membrane dipole movements. The repriming period of charge movements was 100 sec at -100 mV, where the average charge displacement was 29 ± 3 nC$\cdot \mu F^{-1}$. The question then asked was, "To what extent is the maximum available potassium conductance g_K restored during this period at -100 mV and $10°C$?" However it was not possible to separate clearly charge movement and potassium current to see whether their rates of reappearance were identical. Nevertheless maximum conductance after 108 sec repriming at -100 mV was about 0.4 mmho$\cdot \mu F^{-1}$, which was a good deal less than the expected 1–2 mmho$\cdot \mu F^{-1}$. At best the reprimed g_K after 110 sec at -100 mV was only one third to one half of its normal value, at which time charge movement appeared to be complete.

Furthermore 2 mM tetracaine displaced the activation curve of the potassium current[21] by about 25 mV and that of the charge distribution function by only 5 mV, suggesting that the two were not directly related. The rise of potassium conductance was displaced beyond 0 mV toward internal positive potentials by addition of 2 mM tetracaine in solution containing 5 mM Rb_2SO_4 and Na_2SO_4.[22] When the membrane potential was then displaced from holding potentials of -80 and -20 mV to values as positive as $+40$ mV, in the former case the charge movement preceded the development of potassium current as already described[21]; in the latter case, no detectable movement of charge was seen before the rise in potassium conductance. Apparently charge movement was inactivated when fibers were held at -20 mV; but in the absence of such movement, K conductance could still increase by the same amount as in normal fibers. It was therefore possible to get opening of K channels without detectable charge movement. The association of some component of charge movement to initiation of contraction is intriguing, especially as the solution used contained calcium ions.

POTASSIUM CURRENT IN THE REPOLARIZATION OF CARDIAC MUSCLE

The voltage-dependent ionic permeability changes in cardiac muscle appear to be more numerous and complicated than in skeletal muscles. In addition to potassium and sodium fluxes, electrogenic movements of calcium and chloride also seem to take place during the action potential. At least six types of electrically excitable ionic channels may be present in this muscle.

Attempts[23] to quantify potassium movements using ^{42}K efflux and fractionating the effluent from isolated turtle hearts into 200 samples during a single action potential have had limited success because of the distortion of the efflux pattern brought about by the mechanical events[24] in the muscle. The syncytial structure of the cardiac tissue and the tight packing of the cells has limited the reliability of radioisotope technique, as most of the exchange seems to be confined to the extracellular space (paracellular space) immediately adjacent to the myocardial cell which changed in volume during contraction–relaxation. The trabeculum or heart muscle bundle (diameter 300–700 μm), which may contain several thousand cells of about 14 μm in diameter, is enclosed in a thin sheath of endothelial cells which may constitute a barrier to diffusion of ions. In this case changes in potassium concentration within this restricted diffusion space may be sufficient to influence both membrane potential and permeability of the myocardial cell. By the use of double-barreled K-selective microelectrodes it has been possible[25] to monitor the changes in K activity within such spaces during either a single action potential or a series of action potentials of different frequency and relate membrane potential to the accumulation of potassium. The time constant of the electrode to iontophoretically injected KCl was 10 msec[26]; therefore only events lasting longer than 50 msec were examined.

The cardiac action potential in general lasts for several hundred milliseconds and has a slower rate of repolarization than skeletal muscle. Therefore the resolution of potassium currents and of the changes in a_K^0 by means of the K-selective electrode is thereby facilitated. Much interest has centered around the ionic fluxes that take place during the plateau, which in frog muscle lasts for over 300 msec. During this time membrane impedance seemed to be high. The duration of the plateau has been shortened dramatically (Fig. 4.5) by infusion of K-rich solutions through

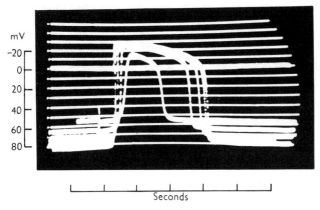

Figure 4.5 Resting and action potentials recorded from the same cardiac muscle fiber. Effect of perfusion of coronary circulation by solutions of different potassium concentration. The steady-state K concentration was 2.7 mM for the longest and about 20 mM for the shortest action potential. From S. Weidmann, *J. Physiol.* **132**, 157 (1956).

the coronary artery. Acetylcholine, which had a similar effect in frog auricular fibers, also increased the potassium conductance in these fibers and also the ^{42}K influx and efflux in frog sinus venosus.[27] It therefore seemed that increase in potassium permeability might contribute to the shortening of the action potential under the conditions described.

It was suggested that during the plateau in the normal cardiac action potential, a gradual accumulation in an extracellular space of restricted diffusion might eventually lead to increased K conductance and the rapid phase of repolarization.[28] Such potassium accumulation was detected with the K electrode. During stimulation of frog ventricular strips at high frequencies[25] (60–90 stimuli·min^{-1}), depolarization of the ventricle reached 20 mV in some cases, while the "extracellular" potassium (Fig. 4.6) measured by the K-selective microelectrode rose from about 3 to 9–12 m.mole·l^{-1} and the action potential shortened. Calcium-free Ringer was used in some cases to eliminate the contraction of the muscle. Following cessation of the stimulation, extracellular potassium concentration decreased to its original value within 40 sec. The rates of buildup and decay of the external potassium here was twice as great in thin as in thick strips, and cooling of the tissue resulted in much larger accumulation of potassium, due presumably to inhibition of its uptake. In voltage clamp experiments employing a single sucrose gap,[29] the duration of the depolarization in the plateau range seemed to determine the level of accumulation of external potassium and the magnitude of the depolarizing afterpotential on removal of the clamp.

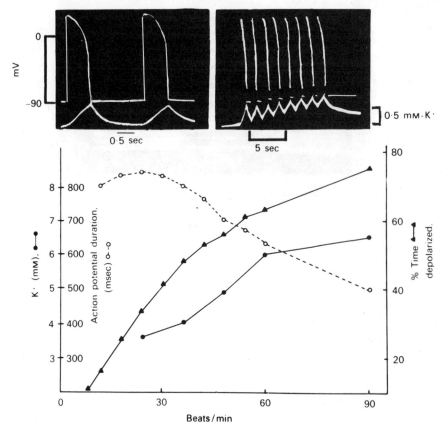

Figure 4.6 Relation between duration of depolarization, K^+ accumulation, and frequency of stimulation. The frequency dependence of action potential duration (open circles), K accumulation (filled circles), and percent of time that membrane is depolarized (filled triangles) are represented. Shortening of AP appears to follow increased $[K]_0$ for high frequencies. K^+ accumulation varies with percent time of depolarization. Upper panel shows APs (upper trace) and K_0 (lower trace). Temperature 22°C, $(Ca^{2+})_0 = 0.2$ mM. From R. P. Kline & M. Morad, *J. Physiol.* **280**, 537 (1978).

In spite of the observed shortening of the action potential following the increase in potassium concentration of the perfusion fluid[28] and the apparent linear dependence of action potential duration in the present experiment on the local potassium concentration prior to the upstroke of the action potential, there was a marked absence of correlation between the time-dependent K current in the positive potential range and the increase in external potassium concentration from 3 mM to 6 mM.[29] A rapid shortening of action potential to a near-steady state also took place in some cases within two beats, whereas a much longer period was required for potassium to accumulate externally to a steady-state level. It was concluded that the delayed current was unaffected by increase in K_0 but that the early (inward rectifying) current might be shifted along the voltage axis as determined by the Nernst equation and that this change might be responsible for the shortening of the action potential.

The form of the action potential is different in the atrium muscle and pacemaker cells of the heart. Because the action potentials differ in amplitude, duration, and shape, it is to be expected that the currents responsible for depolarization and repolarization will also be different. It is difficult to generalize about the electrogenesis in this tissue. For example, in ventricular muscle there was no evidence of the potassium current (I_{K_2}) responsible for the pacemaker systems found elsewhere in the heart.[30]

As in skeletal muscle, the rapid depolarization is brought about by the opening of sodium channels, and these can be blocked by tetrodotoxin. After their removal, depolarizing current will still elicit a small and slower inward current which is suppressed by 2–4 mM manganese or by removal of both sodium and calcium from the medium. This is the inward calcium current through channels which also allow sodium entrance. When both the sodium and calcium channels have been blocked, only an outward current remains, and this is carried for the most part by potassium ions. Chloride ions may also contribute to repolarizing current in Purkinje fibers. The net inward chloride current here is called positive dynamic current, or initial outward current.[8]

The existence of the plateau in cardiac muscle suggests that the potassium system here differs from that of the giant nerve axon of squid and that its description would require a major departure from the Hodgkin–Huxley equations. In cardiac muscle, conductance of the membrane decreased on depolarization, which is consistent with the anomalous rectification of the membrane. Again it has been suggested that the pacemaker potentials of spontaneously active Purkinje fibers are due to a time-dependent decrease in K conductance of the membrane after repolarization which occurs in the presence of depolarizing currents that have not been fully inactivated. Before the application of the voltage clamp technique to this tissue, it seemed that some modification of the Hodgkin–Huxley equation would explain its electrogenesis. For example, the assumption of an instantaneous fall in K conductance on depolarization rising again with a time constant about 100 times longer than in nerve was considered an adequate description of the repolarizing process, but the situation seems to involve more complex diffusion currents.

In the application of voltage clamp to Purkinje fibers, two intracellular electrodes were generally used—one to measure membrane potential, the other to inject current from the clamp amplifier. Short fibers were used taking account of the space constant of the cell, as it was necessary that the whole membrane be clamped to the same potential. In the case of the mammalian myocardium, it has been necessary to use a different technique. Here the clamp was applied across a sucrose gap, and one microelectrode was employed to measure membrane potential. The sucrose gap breaks the electrical circuit for ionic flux externally thereby eliminating the action potential since it has a high electrical resistance. The membrane regions on either side of the gap are then electrically connected through the intracellular phase only; so if current is passed between the two ends, it will flow across the membranes in either bath and through the intracellular fluid. In this way current can be applied intracellularly without resorting to microelectrodes. The potential difference between the two regions may thus be compared with the current used to produce the potential drop. A double sucrose gap technique has also been used in cardiac muscle,[31] whereby a small test region of a strip of muscle <1 mm in length was isolated from the rest of the fiber by two sucrose-filled chambers, one on either side. The length of the test region was usually well within the space constant of the cells, that is, less than 0.5 mm; and this region as well as those on either side of the sucrose gaps were bathed in physiological saline. If high-K solution was used to depolarize one end of the preparation, then the potential difference between the middle region and depolarized end gave an estimate of the membrane potential.

In other cases the sucrose gap was used only for the application of current while membrane potential was measured with the microelectrode. A more recent refinement still in the application of the voltage clamp to cells is the so-called patch clamp, in which a small area of 10^{-4}–10^{-5} cm^2 of surface membrane is electrically isolated from the fluid, bathing the remainder of the cell through the application of concentric capillaries to the surface, the inner capillary being filled with normal external fluid and the outer capillary with sucrose solution.

Good potential control and response times can be achieved by these methods. But the nonuniformity of potential can be a major source of error in the application of the voltage clamp to cardiac muscle. Unlike the squid axon, which approximates better to a cylindrical fiber, a strip of cardiac muscle is made up of tightly packed cells with intervening clefts which have a high electrical resistance, so that voltage records are made up of two components—one representing membrane potential and the other the product of clamp current and resistance within the clefts which may be changing with current flow.

In cardiac muscle the rate of repolarization is generally less than 0.1%

of the maximum depolarization rate and is sensitive to factors such as plasma potassium concentration. The marked fall in membrane current that follows the initial rapid repolarization of the membrane may even produce a notch in the plateau of the action potential.

The membrane current changes in relation to time and voltage in cardiac muscle. While the net repolarizing current in the heart tends to be small, this does not necessarily mean that the individual currents carried by sodium, calcium, and potassium are also small. However by measuring the potential changes in response to small pulses of current at different times during the action potential, it was shown that total membrane conductance fell during the plateau.[28] That is to say, the potential change for a small applied current increased during this period, indicating a fall in "slope" conductance dI/dV. The membrane current rises steadily during the plateau, reaching a maximum at the steepest part of the repolarization and falling again toward zero when the membrane is fully repolarized in diastole.

In Purkinje fibers, current was maximal at about -60 mV, falling to a minimum at -20 mV.[32] Voltage clamp experiments in sheep ventricular fibers indicated that normal repolarization arose from a time-dependent decrease in inward current rather than a time-dependent increase in outward current.[33] The inactivation of inward current in this case was slow, while outward potassium current was small. In Purkinje fibers, where inactivation of calcium current appeared to be much faster with a time constant of 50 msec,[34] inactivation of this current could not account for a repolarization in an action potential lasting for over 300 msec. In this case outward potassium current was more important. The low conductance during the plateau in this muscle is apparently brought about by the inward K rectification which brings K conductance below the resting level while at the same time reducing the energy requirement for the production of the action potential, because depolarization is maintained without increase in net influx of sodium and calcium. The inward potassium rectification is said to be time independent since it follows instantaneously on depolarization.

Membrane conductance in Purkinje fibers may be influenced either by a change in K_0 or by current injection. In the former case, g_K was increased[35,36] in conformity with the predictions of the Goldman constant field equation, eq. (17), and ^{42}K efflux was increased likewise, when K_0 was increased. Under these conditions the membrane was depolarized. Current injection, on the other hand, produced anomalous rectification, or a decrease in g_K on depolarization,[37] resulting in an N-shaped current–voltage relation (Fig. 4.7). Thus the two curves relating current to voltage crossed one another. Due to anomalous rectification, P_K declined to about

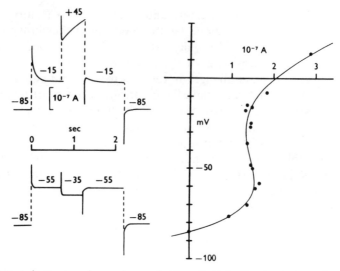

Figure 4.7 Left, Tracing of current recorded in a Purkinje fiber in Na-free solution during double step changes in membrane potential. Resting potential = −85 mV. Potentials during steps indicated by figures above current records. Right, Steady-state I/V relation. Note negativity of slope between −20 and −60 mV. From R. E. McAllister & D. Noble, *J. Physiol.* **186,** 632 (1966).

a tenth of that found at $E_m = E_K$ when the former potential was reduced to −50 mV. The relationship between the potassium conductance g_K here and that occurring at $E_m = E_K$, namely $(g_K)_{E_K}$, was described by the equation

$$g_{K_1} = (g_K)_{E_K} F_1 (E_m - E_K) \tag{30}$$

where the term F_1 describes the anomalous rectification. When this equation was applied to experimental data in addition to the voltage-dependent anomalous conductance g_{K_1}, a time-dependent change in potassium conductance g_{K_2}, otherwise known as the delayed rectification, also became evident. For example, while the computed current should have remained virtually constant at voltages more positive than −70 mV, it was found experimentally that at potentials more positive than −30 mV a steadily increasing outward current developed. This outward current called i_x to distinguish it from the pure K current, i_{K_2}, associated with the pacemaker activity was subsequently found[38] to have two components, i_{x_1} and i_{x_2}, each showing exponential changes and obeying kinetics of the Hodgkin–Huxley type. The first of these (Fig. 4.8) was activated with a time constant of about 0.5 sec at the plateau potential. At more positive and negative potentials, its time constant was shorter. The second com-

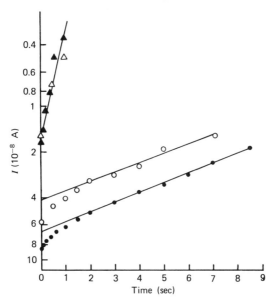

Figure 4.8 Analysis of time course of current activated by depolarization to −10 mV. Slow current change during depolarization was measured as deviation of current from steady-state value (closed circles). Slow component fitted by a straight line with intercept 6.6×10^{-8} Å and time constant 5.25 sec was subtracted from these points to give fast component (closed triangles) which was fitted by a straight line with intercept 2.3×10^{-8} Å and time constant 0.55 sec. Open circles were obtained from tail currents. From D. Noble & R. W. Tsien, *J. Physiol.* **200**, 205 (1969).

ponent was activated extremely slowly, with a time constant at plateau of 4 sec and so was unlikely to contribute to the action potential.

The first of these currents, which probably determined the duration of the plateau and its termination, has been studied[39] by clamping the membrane of Purkinje fibers at −30 mV, which is about 10 mV beyond the threshold for its activation. The potential was then changed from this holding voltage by depolarizing and hyperpolarizing current pulses of different magnitude. The former produced an outward current flow, but the latter tended to switch off current. The reversal potential for the outward current was at −80 mV, which was about 20 mV less negative than E_K. It was therefore suggested that the channels carrying this current showed incomplete selectivity for potassium ions, allowing some sodium ions also to pass through.

The time-dependent component of late outward potassium current was then found to have a very negative threshold in addition to that at −40 mV and associated with the plateau. The former was activated between −60 and -90 mV and was related to the pacemaker potential. It appeared[30] as an outward declining tail current on repolarizing to holding potentials positive to its reversal potential, which was given by E_K.[40,41] At still larger hyperpolarizations, the current reversed, appearing as a declining inward current. The decline in conductance g_{K_2} was seen[42] when after repolarization of the action potential (during which the current was activated) the membrane potential was clamped at the maximum diastolic value. In this case membrane current was zero at first, but an inward current developed with a time constant of about 3 sec. This was due to the inactivation of outward current i_{K_2} in the presence of incompletely inactivated inward currents carried perhaps by sodium. This late time-dependent outward potassium current seems to be responsible for the transient hyperpolarization that follows the action potential. This current however showed inward rectification when the membrane was depolarized rapidly from -90 mV to -50 mV or less, so it could not contribute much to repolarization during the plateau. This current was not present in the myocardium. Its inactivation produced[43] the prepotential of pacemaker cells and the drift of membrane potential toward the firing level, and a rise in membrane resistance. While pacemaker activity in Purkinje fibers may be adequately explained by a fall in g_{K_2} with time in the membrane potential range of -90 to -60 mV, pacemaker activity at the sinoatrial node of rabbit showed characteristics which were not in keeping with this mode of electrogenesis. In the latter, the slow depolarization to the firing level was accompanied by a fall in membrane resistance,[44] suggesting a rise in sodium conductance rather than a fall in potassium conductance. It was also activated over a less negative voltage range, -40 mV to -60 mV.[45]

The relative importance of time-dependent and voltage-dependent outward currents in the repolarization process is a matter of some controversy, with some claiming a more important role for time-independent or background K current in the process. The delayed rectification, which seemed to be responsible for the termination of the plateau,[39,41] was not seen in all preparations of Purkinje fibers; thus other conductance changes were indicated.

It was found[46,47] that the membrane current at 0 mV in cat ventricular fibers could be accounted for by three time-dependent components: (a) the inactivation of an inward current through calcium channels (also used by sodium ions), with a time constant of 90 msec; (b) the activation of an outward potassium current I_K, with a time constant of 370 msec; and (c) the activation of a slower less specific current I_x, similar to that described above but with a time constant of 3 sec. These time constants were also voltage dependent. The voltage dependence of the time constant for I_K was bell shaped (Fig. 4.9), being about 150 msec at -90 mV, 500 msec at

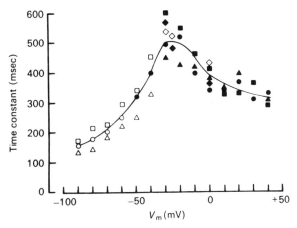

Figure 4.9 Voltage dependence of the I_K time constant in cat ventricle. From T. F. McDonald & W. Trautwein, *J. Physiol.* **274**, 217 (1978).

−25 mV, and 300 msec at +30 mV. Naturally there was no pacemaker conductance in this tissue. The determination of the reversal potential of I_K was complicated by the accumulation and depletion of external potassium.

In Purkinje fibers the steady-state potassium conductances g_K and g_{K_2} were found to be inhibited when 20 mM cesium ion was added to the Tyrode solution.[48] Here the membrane potential fell to about −60 mV within 1 min. In the presence of this cation, which has been shown to interfere with potassium exchange across skeletal muscle membranes,[49] the steady-state current was linearly related to the applied clamp voltage, indicating the absence of inward rectification. It was also insensitive to changes in $[K]_0$. The use of Cs appeared to uncover another membrane current which had apparently been overlooked because of a preoccupation with passive membrane currents. The Cs-sensitive membrane current of Purkinje fibers in low-K could not be accounted for in terms of passive potassium efflux, and so it was suggested that this cation might also activate an electrogenic sodium pump in the membrane.

Under control conditions at −65 mV, membrane current was about 25 nA and outward, whereas after exposure to cesium it became 4 nA in the opposite direction. When cesium concentrations down to 0.1 mM were used, the pacemaker current i_{K_2} was reduced selectively while the instantaneous outward current i_{K_1} was unaffected. The effect of caesium ions was not on the activation and deactivation variable since the use of stronger depolarizing currents was without effect on the outward K currents. It seemed therefore that the conductance parameter $\bar{g}_{K_2}$ was being blocked by Cs. In Fig. 4.10 the changes in steady-state current–voltage relation induced by 20 mM Cs may be seen. Here both i_{K_1} and i_{K_2} were reduced by Cs and the N-shaped curve so typical of inward rectification became linear. The reversal potential of the Cs-sensitive component was between −80 and −85 mV. By subtraction of the curves in the presence and absence of cesium the Cs-sensitive current (dashed line) was obtained. It is also possible to derive the conductance changes Δg controlling the Cs-sensitive current by dividing the change in current ΔI by the force driving ions through the membrane, namely, the clamp voltage minus the reversal potential. This indicated that at clamp voltages negative to the reversal potential (hyperpolarizing) conductance was large, decreasing rapidly as the clamp voltage approached, and became positive to, the reversal potential. All the evidence strongly indicated that cesium blocked potassium conductance. On the other hand the potassium currents in Purkinje fibers appeared to be increased on injection of calcium ions into the cell.[50]

The cesium appeared however not only to block the inward rectifying potassium channels but also produced an outward current which was sensitive to dihydro-ouabain and was therefore perhaps produced by the electrogenic pumping of sodium from the cell. The existence of such currents had already been indicated by earlier experiments in which the outward steady-state current of Purkinje fibers was reduced by 10–20 nA within minutes of adding dihydro-ouabain to the bathing fluid,[51] and this decrease did not depend on membrane potential. The increase in sodium pump current in response to addition of cesium was found to be greater with lower $[K]_0$, being about 30 nA in 1 mM K_0 and falling to near zero at 5.4 mMK. This was probably due to change in the coupling ratio of Na/K due to interaction of Cs with K sites.

GATING AND THE EFFECT OF CALCIUM AND HYDROGEN IONS

The view that part of the electric field within the membrane responsible for the gating of ionic channels depends on surface negativity has recently found support[52] in the effects of calcium ions and protons on sodium and potassium conductances measured by two microelectrode voltage clamps in sheep Purkinje fibers. In these studies the K currents were more amenable to measurement than those of sodium. The full activation curve of the pacemaker K current and not just threshold was shifted in a positive direction by increase in external calcium concentration. In a semilog plot a shift of 8 mV per fourfold change in $[Ca]_0$ was obtained, and this was restricted to the voltage dependence of the activation curve, the absolute

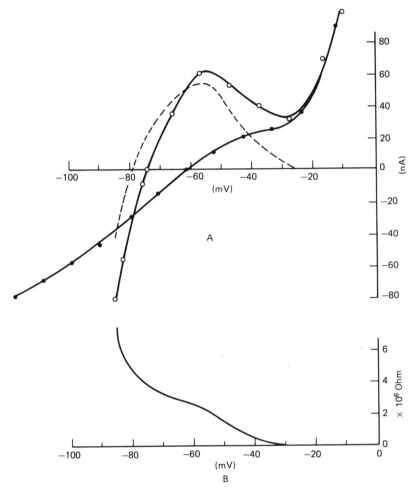

Figure 4.10 Changes in the steady-state *I/V* relation induced by 20 m*M* Cs. Membrane current flowing at clamp step (2 sec duration) plotted against clamp voltage. The Tyrode solution in which Purkinje fibers were bathed contained 5.4 m*M* K. (A) "Cs-sensitive current" (broken line) was obtained by subtracting current two minutes after exposure to Cs (closed circles) from control current (open circles). (B) Changes in membrane conductance induced by Cs. The Cs-sensitive current of (A) was divided by driving force (clamp voltage minus reversal potential). From G. Isenberg, *Pflügers Arch.* **365,** 99 (1976).

amplitude of i_K^2 being not significantly changed.[53] The results of titrations were consistent with a surface potential of -18 mV under normal conditions. The addition of calcium ions which shield or bind the external negative sites and of protons which also reduce the magnitude of i_K^2 when fully activated may be of clinical importance. The binding of calcium might be expected to have a hyperpolarizing effect on the actual transmembrane potential without affecting the potential measured by electrodes in the bulk solutions of either side of the membrane. The results obtained with protons were less clear than those with calcium, and this was probably due to the former cation entering the fibers and affecting both surfaces of the membrane.

The effects of low pH and of calcium ions on the steady-state currents measured over a wide range of potential in Purkinje fibers were consistent with the accumulation of potassium ions in clefts of the fibers mediating the effects observed. The effects of acidosis were prevented by increasing $[K]_0$ from its normal range up to 10.8 mM. On exposure to pH 6.6 (reduction of HCO_3^-) at 5.4 mM K_0 there was marked depolarization of the membrane; and at E_m more negative than -10 mV, membrane current was in an inward direction. After prolonged exposure to acid solution, the reversal potential of the current i_K^2 moved from -105 mV at pH 7.2 to about -84 to -72 mV, due presumably to an increase in extracellular potassium in the clefts of the tissue. The effect of acidosis at normal $[K]_0$ was like that produced by pump inhibition in the presence of toxic doses of cardiac glycoside.

CONCLUSIONS

Some uncertainties remain concerning the role of potassium ions in the repolarization of cardiac muscles. However, one promising approach in this field is the use of K-selective microelectrodes to measure changes in the clefts of the muscle fibers.[54,55] This may help clarify the nature of the valvelike regulation of potassium movements and of the positivity of some reversal potentials relative to "apparent" E_K values. Another approach involves the use of chemical tools such as cesium and lithium. The first has proved effective in resolving potassium currents, while the second may help delineate the involvement of electrogenic sodium pumping in the process of repolarization.

In connection with the former approach and the restricted diffusion immediately outside cardiac cell tissue, the following should be mentioned. The application of hyperpolarizing currents to ventricular trabeculae or papillary muscle of cat and sheep produced an inward current which was dependent on external potassium concentration and declined with time. When the hyperpolarization had been prolonged, the membrane current tended to be outward on returning to the resting potential and the membrane potential had become more negative. The kinetics of decline of inward current under hyperpolarizing clamp coupled with those of the outward current after prolonged hyperpolarization and the resulting increase in membrane potential were attributed[54] to depletion of potassium ions and their subsequent replenishment in an extracellular space of restricted diffusion, occupying about 3% of total muscle volume, and not to a time-dependent change in membrane permeability.[56] Such space might explain the valvelike regulation of potassium movement across the cell membrane or the phenomenon of inward rectification.

REFERENCES

1. A. L. Hodgkin, A. F. Huxley & B. Katz, *Arch. Sci. Physiol.* **3**, 129 (1949).
2. A. L. Hodgkin & A. F. Huxley, *J. Physiol.* **117**, 500 (1952).
3. G. Burnstock, M. E. Holman & C. L. Prosser, *Physiol. Rev.* **43**, 482 (1963).
4. G. W. Beeler & H. Reuter, *J. Physiol.* **207**, 191 (1970).
5. R. H. Adrian, W. K. Chandler & A. L. Hodgkin, *J. Gen. Physiol.* **51**, 188s (1968).
6. R. H. Adrian, W. K. Chandler & A. L. Hodgkin, *J. Physiol.* **208**, 607 (1970).
7. R. J. Lipicky & S. H. Bryant, *J. Gen. Physiol.* **50**, 89 (1966).
8. J. Dudel, K. Peper, R. Rüdel & W. Trautwein, *Pflügers Arch.* **295**, 197 (1967).
9. R. J. Lipicky, S. H. Bryant & J. H. Salman, *J. Clin. Invest.* **50**, 2091 (1967).
10. R. H. Adrian & S. H. Bryant, *J. Physiol.* **240**, 505 (1974).
11. R. H. Adrian & M. W. Marshall, *J. Physiol.* **258**, 125 (1976).
12. B. Frankenhaeuser & A. L. Hodgkin, *J. Physiol.* **131**, 341 (1956).
13. L. D. Peachey, *J. Cell. Biol.* **35**, 209 (1965).
14. S. Nakajima, S. Iwasaki & K. Obata, *J. Gen. Physiol.* **46**, 97 (1962).
15. R. H. Adrian, W. K. Chandler & A. L. Hodgkin, *J. Physiol.* **208**, 645 (1970).
16. R. H. Adrian & L. D. Peachey, *J. Physiol.* **235**, 103 (1973).
17. A. Duval & C. Leoty, *J. Physiol.* **278**, 403 (1978).
18. R. D. Keynes & E. Rojas, *J. Physiol.* **239**, 393 (1974).
19. M. F. Schneider & W. K. Chandler, *Nature* (Lond.) **242**, 244 (1973).
20. R. H. Adrian & A. R. Peres, *Nature* (Lond.) **267**, 800 (1977).
21. W. Almers, *J. Physiol.* **262**, 613 (1976).
22. R. H. Adrian & R. F. Rakowski, *J. Physiol.* **278**, 533 (1978).
23. J. M. O'Brien & W. S. Wilde, *Science* **116**, 193 (1952).

24. J. F. Lamb & J. A. S. McGuigan, *J. Physiol.* **195**, 283 (1968).
25. R. P. Kline & M. Morad, *J. Physiol.* **280**, 537 (1978).
26. E. Neher & H. D. Lux, *J. Gen. Physiol.* **61**, 385 (1973).
27. W. Trautwein & J. Dudel, *Pflügers Arch.* **266**, 324 (1958).
28. S. Weidman, *J. Physiol.* **132**, 157 (1956).
29. L. Cleeman & M. Morad, *Science* (N.Y.) **191**, 90 (1976).
30. K. A. Deck & W. Trautwein, *Pflügers Arch.* **280**, 63 (1964).
31. O. Rougier, G. Vassort & R. Stämpfli, *Pflügers Arch.* **301**, 91 (1968).
32. D. Noble & R. W. Tsien, The repolarization process of heart cells in *Electrical Phenomena in the Heart*, W. C. deMello, Ed., Acad Press, New York, 1972, pp. 133–161.
33. G. Giebisch & S. Weidmann, *J. Gen. Physiol.* **57**, 290 (1971).
34. M. Vitek & W. Trautwein, *Pflügers Arch.* **323**, 204 (1971).
35. E. E. Carmeliet, in *Chloride and Potassium Permeability in Cardiac Purkinje Fibres*, Presses Académiques Européennes Société Cooperative, Brussels, 1961.
36. A. E. Hall, O. F. Hutter & D. Noble, *J. Physiol.* **166**, 225 (1963).
37. R. E. McAllister & D. Noble, *J. Physiol.* **186**, 632 (1966).
38. D. Noble & R. W. Tsien, *J. Physiol.* **200**, 233 (1967).
39. D. Noble & R. W. Tsien, *J. Physiol.* **200**, 205 (1967).
40. K. Peper & W. Trautwein, *Pflügers Arch.* **309**, 356 (1969).
41. D. Noble & R. W. Tsien, *J. Physiol.* **195**, 185 (1968).
42. M. Vasalle, *Am. J. Physiol.* **210**, 1335 (1966).
43. J. Dudel, K. Peper, R. Rüdel & W. Trautwein, *Pflügers Arch.* **296**, 308 (1967).
44. I. Seyama, *J. Physiol.* **255**, 379 (1976).
45. O. F. Hutter & W. Trautwein, *J. Gen. Physiol.* **39**, 715 (1956).
46. T. F. McDonald & W. Trautwein, *J. Physiol.* **274**, 193 (1978).
47. T. F. McDonald & W. Trautwein, *J. Physiol.* **274**, 217 (1978).
48. G. Isenberg, *Pflügers Arch.* **365**, 99 (1976).
49. L. A. Beaugé, A. Medici & R. A. Sjödin, *J. Physiol.* **228**, 1 (1973).
50. G. Isenberg, *Pflügers Arch.* **371**, 77 (1977).
51. G. Isenberg & W. Trautwein, *Pflügers Arch.* **350**, 41 (1974).
52. R. H. Brown Jr. & D. Noble, *J. Physiol.* **282**, 333 (1978).
53. R. H. Brown Jr, I. Cohen & D. Noble, *J. Physiol.* **282**, 345 (1978).
54. D. W. Maughan, *J. Membr. Biol.* **28**, 241 (1976).
55. J. J. Van der Walt & E. E. Carmeliet, *Arch. Int. Physiol. Biochem.* **79**, 149 (1971).
56. G. F. Wooten, D. H. Park, T. H. Joh & D. J. Reis, *Nature* (Lond.) **275**, 322 (1978).

5
Active Transport of Potassium in Nonepithelial Cells

The active transport of potassium cannot be considered in isolation from that of other cations across the cell membrane, because its net movement into the cell is usually coupled to the excretion of sodium ions or hydrogen ions, the latter being usually against a greater electrochemical gradient. In skeletal muscles at rest, the potassium ions are virtually at the same electrochemical potential inside and outside the fiber and the membrane is quite permeable to this cation, so the need for its active uptake is not immediately obvious. In hepatocytes[1] and erythrocytes,[2] on the other hand, measured membrane potentials were only −44 mV and −14 mV, respectively, whereas the distribution of potassium ions indicated an E_K value of about −80 to −90 mV. In this case work needed to be done to move potassium ions into the cells.

The erythrocyte has proved to be a particularly convenient cell in which to study active transport because its internal milieu can be varied at will. The cells were made to swell by suspension in slightly hypotonic solution, becoming just sufficiently leaky to lose their hemoglobin and some internal proteins and metabolites. Such cells were then reconstituted[3] by being brought back to isotonicity and incubated at 37°C. Before resealing,

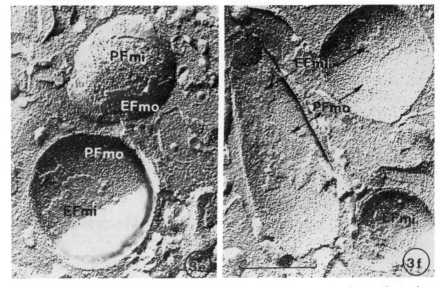

Plate 6 Freeze-fracture profiles of mitochondria from intact hepatocytes showing faces of (a) inner (mi) and (b) outer (mo) membranes. PF and EF represent protoplasmic face and external faces, respectively. From G. A. Losa, E. R. Weibel & R. P. Bolender, *J. Cell Biol.* **78**, 289 (1978).

the cells could be equilibrated with a solution which was to become its internal environment. The so-called ghost cells prepared in this way have been used extensively to study the properties of membrane-bound (Na^+ + K^+)-dependent adenosine triphosphatase (ATPase), believed to be responsible for the active transport of these cations. This enzyme system appears to be an integral protein[4] penetrating the lipid bilayer of the membrane (see Plates 6 and 7), and because of its lipid solubility it is only isolated with great difficulty by the use of detergents.[5]

GENERAL STRUCTURE OF THE TRANSPORT PROTEIN

The transport enzyme has been purified[6] in membranes from the renal medulla by the selective extraction of other membrane proteins with sodium dodecyl sulfate, leaving the former intact and functional. The (Na^+ + K^+)-ATPase appeared to be made up of two polypeptide components, a catalytic protein of MW 100,000 comprising 60–70% of the total protein and a smaller sialoglycoprotein.[7,8] The very hydrophobic light-

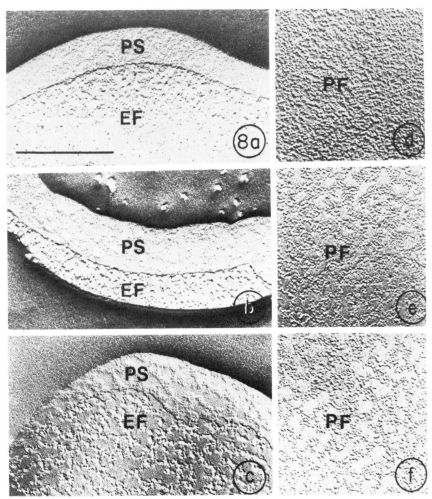

Plate 7 Freeze-etch electron micrographs (×60,000) showing (a) the protoplasmic surface (PS) of the erythrocyte membrane and extracellular face (EF) of fresh ghosts; (b) the protoplasmic surface of ghosts pretreated in 0.5 mM sodium azide–phosphate buffer, pH 9 for 10 hr at 37°C; and (c) of inside-out vesicles prepared by incubating fresh ghosts in 0.5 mM Na_2HPO_4, pH 8.2 for 24 hr at 10°C. For comparison, the corresponding protoplasmic faces (d, e, and f) are shown. Bar, 5000 Å. From D. Shotton, K. Thompson, L. Wofsy & D. Branton, *J. Cell Biol.* **76**, 512 (1978).

sensitive compound 5-^{125}I-iodonaphthyl-l-azide (INA)[9,10] (0.5 mole per mole of enzyme) seemed to label that part of the pump molecule in contact with the lipid phase of the membrane; and after exposure to this compound, a segment of about MW 12,000 from the larger polypeptide chain was found to combine with the label. Graded proteolysis has helped the identification of the position of this segment within the chain. After extensive proteolysis only 45% of the original protein remained in the membrane, but less than 10% of the radioactivity had been released. As only about 20% of the total radioactivity seemed to be free in the lipid phase of the membrane, this suggested that the bulk of the activity was in the interface between proteins and lipid. The ATPase activity was uninhibited by the presence of INA unless its concentration was increased 50- to 100-fold, at which level incorporation of INA into the glycoprotein was also seen.

It should be said that all catalytic activity was destroyed by limited proteolysis[12] whereas exhaustive proteolysis did not release the bulk of the label. It was found that INA did not interfere with binding of ATP to the enzyme system although at high concentration it might inhibit catalysis by perturbing protein–lipid interactions. When the labeling of the large protein was compared in sodium- and in potassium-containing media, there was a 10–25% greater incorporation of the probe into the large chain in the presence of potassium, which would be compatible with a conformational change,[13] exposing more hydrophobic amino acid side chains to the reactive hydrophobic compound within the lipid bilayer.

While the ionic environment of the membrane proteins seemed capable of changing the contact between lipid and protein, the integrity of the membrane phospholipid was also important for the normal (Na^+ + K^+)-ATPase activity of the protein[14–16] and therefore for its interaction with the monovalent cations. For example, treatment of red blood cell membranes with pure phospholipase C inactivated this enzyme while increasing the K-dependent phosphatase activity of the membrane protein.[17] The standard substrate used for the measurement of K-dependent phosphatase activity was p-nitrophenyl phosphate (P-NPP), and its maximum rate of hydrolysis and maximum activating effect of potassium were both increased under these conditions. It has been suggested[18] that the K-activated phosphatase activity present in cell membranes is a property of the Na–K pump; its relationship to the pump mechanism is discussed later. While the (Na^+ + K^+)ATP-ase activity is inhibited by ouabain, the K-dependent phosphatase activity is unaffected by this cardiac glycoside but is sensitive to the SH reagent N-ethylmaleimide (NEM).

THE SODIUM-POTASSIUM CARRIER OF RED CELL "GHOSTS"

The membrane (Na^+ + K^+)-dependent ATPase system which was first investigated[19] in the form of subcellular particles or vesicles (see Plate 8) hydrolyzed adenosine triphosphate to adenosine diphosphate and inorganic phosphate in the presence of sodium and potassium. The nature of this reaction, coupled with its inhibition by ouabain[20] or strophanthidin, which also blocked transport of sodium and potassium across cell membranes, led to its identification as the monovalent cation pump of the membrane. Active transport by definition is a process requiring expenditure of metabolic energy; therefore the availability of energy in the hydrolysis of ATP, coupled with the observation that the cations required for activation of the enzyme system were those transported, provided a logical basis for a membrane transport system. Through the use of the ghost cells, the vectorial properties of this system soon became evident[21,22] when it was found that sodium ions activated it internally but not externally and potassium ions were needed outside but not inside the cells. The ATP was supplied internally as already described and three moles of sodium were excreted and two moles of potassium taken up for each mole of ATP hydrolyzed. The cation transport and ATP hydrolysis seemed to be linked in two stages.[23] Sodium was required for the phosphorylation of a membrane protein by ATP with the release of ADP. Potassium ions were required for the hydrolysis of the phosphoprotein and release of inorganic phosphate, P_i.

The "partial reactions" of the sodium pump were examined in detail[24] in resealed red cell ghosts in relation to the particular ion transported and to its direction of movement. It was discovered,[25] for example, that the pump as a whole was reversible, so that if the electrochemical gradient against which it operated was increased while the energy available to the pump, that is, the [ATP]/[ADP] ratio was reduced within the cells, ATP could be synthesized through a downhill movement of sodium and potassium ions. For example, when $[Na]_i$ was low and $[K]_i$ very high and ^{32}P-labeled orthophosphate was introduced into the ghost cells, which were then suspended in K-free high-Na medium, the incorporation of the label into ATP took place, and this process could be blocked by ouabain. The incorporation was also demonstrated in intact prestarved human erythrocytes.

Having established the reversibility of the pump as a whole, the individual steps of the process were examined. A ouabain-sensitive Na-Na exchange which required the presence of ATP and ADP within the cells but which did not result in net hydrolysis of ATP was observed[26,27] in the

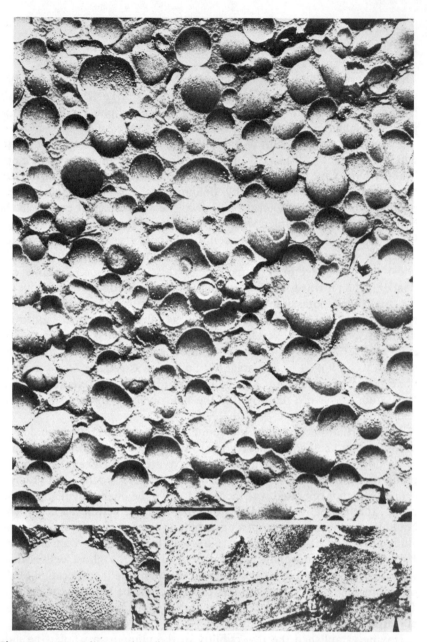

Plate 8 (a) Freeze fracture of a pellet consisting of a homogeneous population of vesicles. Bar 1 μm, ×70,000. (b) As above but showing a vesicle which bears large aggregates of particles. Bar 0.5 μm, ×60,000. (c) Freeze-etched preparation fixed in suspension. A bundle of fibrils may be seen among the vesicles, in the ice, not in the membrane fracture plane (arrow). Bar 0.1 μm, ×138,000. From A. Monneron & J. d'Alayer, *J. Cell Biol.* **77,** 211 (1978).

absence of external potassium. In the normal mode of action, the pump did not require both ATP and ADP but only the former nucleotide. This partial reaction was best studied by the use of different sodium labels, for example, ^{22}Na within and ^{24}Na outside the ghost cells, and expressed by the equation

$$\text{ATP} + \text{enzyme} \xrightleftharpoons[^{22}\text{Na out}]{^{24}\text{Na in}} \text{ADP} + \text{enzyme-P} \qquad (31)$$

Magnesium ions were also required in this reaction, but it was usually combined with the nucleotides. At higher concentrations magnesium also appeared to play a part in conformational changes in the phosphorylated protein, preventing the dephosphorylation of the protein by reversal of the above reaction. That is to say, it was not possible to demonstrate the incorporation of labeled ADP into ATP unless very low levels of Mg^{2+} were present in the cells or unless they were pretreated with NEM or with an equimolar mixture of 2,3-dimercaptopropanol (BAL) and arsenite. It seemed necessary then to add to the above reaction a new step, namely, a Mg-dependent conversion of the phosphorylated protein to another form, *enzyme-P, this reaction being virtually irreversible. The overall mechanisms catalyzed by the three cations might then be summarized in the following reactions:

$$\text{ATP} + \text{enzyme} \xrightleftharpoons{\text{Na}^+ + \text{Mg}^{++}} \text{ADP} + \text{enzyme-P} \qquad (32)$$

$$\text{NEM, BAL- arsenite} \updownarrow \text{Mg}^{++}$$

$$\text{enzyme} + P_i \xrightleftharpoons{\text{K}^+} \text{*enzyme-P}$$

In the cells pretreated with BAL–arsenite, the conversion of enzyme-P to *enzyme-P did not take place, but the former accumulated and could therefore react with radioactive ADP to produce labeled ATP. The validity of this scheme was supported by the finding that in preparations pretreated with BAL–arsenite and exposed to γ ^{32}P-ATP, in the presence of Na$^+$ and Mg^{2+}, the membranes became labeled with radiophosphate, but this activity was not released in the presence of potassium ions, indicating an absence of *enzyme-^{32}P. Furthermore when membranes were labeled with ^{32}P from ^{32}P-ATP in the presence of Na$^+$ and Mg^{2+} and a large amount of unlabeled ATP was then added, release of label gave an indication of breakdown of phosphorylated intermediates. In membranes pretreated with BAL–arsenite, addition of ADP but not of K$^+$ promoted release of label, apparently through breakdown of enzyme-P, whereas in untreated membranes it was the K$^+$ rather than ADP which was effective in this respect, suggesting the breakdown of *enzyme-P.

The Na–Na exchange varied with the concentrations of ADP within the ghosts, being relatively large when the latter was high. It seemed likely therefore that the ion exchange was brought about by reversal of membrane phosphorylation by ATP, eq. (31), right to left; that is to say, Na efflux was associated with phosphorylation of enzyme by ATP and sodium influx with the reverse reaction. Unfortunately this simple model was put in doubt by the observation that oligomycin inhibited Na–Na exchange[28] but stimulated ^{14}C-ADP–ATP exchange.[29] It was necessary therefore to propose a looser coupling with an intermediate reaction which was oligomycin sensitive or to assume that both reactions were independent and that ADP facilitated Na–Na exchange by keeping the concentration of phosphorylated protein of the membrane below a critical value at which it became inhibitory to this exchange. At any rate, the ouabain-sensitive Na–Na exchange and Na–K exchange appeared to use the same carrier sites since both were blocked by antisera prepared in rabbits by the injection of partially purified (Na$^+$ + K$^+$)-ATPase from pig kidney.[30]

POTASSIUM–POTASSIUM EXCHANGE AND DEPHOSPHORYLATION

A later stage in the operation of the sodium pump is the dephosphorylation of the enzyme system with the release of inorganic phosphate. This is the reaction that requires the presence of potassium ions. The finding[31] of a small ouabain-sensitive efflux of potassium from starved red cells indicated the reversibility of this process also. The efflux was also blocked by the antisera to the (Na$^+$ + K$^+$)-ATPase and was reduced by 23% by removal of external potassium. The most suggestive property of the potassium efflux was its dependence on a relatively high level of inorganic phosphate within the cell. The K–K exchange rate measured with radioisotopes reached its half-maximum at $1.7 \times 10^{-3}\,M\ P_i$. When inosine was introduced into the medium,[31] it entered the cells and was split phosphorolytically into ribose-1-phosphate and hypoxanthine, and simultaneously [P]$_i$ of the cytoplasm was reduced from 3.3×10^{-3} to $7.8 \times 10^{-5}\,M$ over 15 min. This led to inhibition of potassium efflux. The inference from this finding was that the binding and carriage of potassium out of the cells was coupled with the phosphorylation of the transport enzyme with P_i in a reversal of the normal mode of operation of the pump, eq. (32).

Reduction of the intracellular sodium concentration by attenuating the Na–K exchange favored the K–K exchange. The very strong inhibitory effect of cell sodium could be seen by replacing most of the intracellular cation by choline. When $[K]_i$ was about 9 mM, the K–K exchange was inhibited[32] by 90% when internal sodium concentration was raised to 4 mM from a value of 1 mM Na, at which Na–K exchange was virtually absent.

Although it was possible to change $[ADP]_i$ over a fairly wide range without interfering with K–K exchange, there was a strong positive correlation between $[ATP]_i$ and the ouabain-sensitive K efflux rate. This was an unexpected result since ATP was not hydrolyzed to any significant degree. It was suggested however that ATP in combination with the enzyme might act as a carrier for potassium without expenditure of energy.

A striking asymmetry was found in the affinity of the sites on either side of the membrane for potassium ions in the activation of the enzyme and exchange process, but the direction of this asymmetry was the opposite to that of the sodium ions. When red cells were incubated in choline media with different concentrations of potassium present, the oubain-sensitive efflux of ^{42}K reached its half-maximum when $[K]_0$ was about 0.25 mM, while the half-maximum for the sites inside the membrane was at 10 mM K_i.[32] In the case of Na–Na exchange on the other hand, a maximum was reached at $[Na]_i$ of 10–15 mM, but it was linearly related to $[Na]_0$ over a wide range and up to 150 mM at least. This finding of course was in line with the normal mode of operation of the pump, showing its higher affinity for sodium within the cell and for potassium outside. The pitting of the electrochemical potential of the cations across the cell membrane against the chemical potential of phosphate provides an elegant example of energy conservation, similar to that in mitochondria, which is considered later.

When the transport enzyme was solubilized by Lubrol and purified,[33] it still had the same optimum pH, optimum ATP/Mg^{2+} ratio, apparent K_m for ATP, and optimum Na/K ratio as the enzymes in crude membranes. It seems unlikely however that incorporation of ^{32}P into ATP could take place without vectorial cation movements made possible by the use of cell ghosts, although such a claim has been made.[34]

LOW- AND HIGH-POTASSIUM RED BLOOD CELLS

It has been known since the last century[35] that the concentrations of potassium and sodium in red blood cells vary greatly between and within species. Among the herbivores that have been extensively studied in this regard some have an intracellular sodium concentration four to five times the intracellular potassium concentration (Fig. 5.1) in the steady state.[36,37]

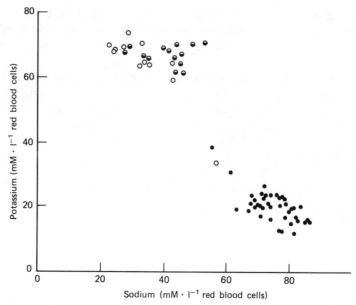

Figure 5.1 Concentrations of potassium and sodium in the red blood cells of the goat and ox: ○ Anglo-Nubian goats, ◐ British Saanen goats, ● cattle. From J. V. Evans & A. T. Phillipson, *J. Physiol.* **139**, 87 (1957).

The uptake rate for potassium and also the level of $(Na^+ + K^+)$-ATPase activity were also found to be four times greater in red cells of the high-potassium (HK) type of sheep than in those of the low-potassium (LK) type.[38] Associated with the difference in electrolyte contents of these cells was a difference in the hemoglobin.

Comparative studies of the HK and LK cell types seemed to offer a unique opportunity for identifying and characterizing the enzyme systems and metabolic processes responsible for cation transport. Perhaps some membrane proteins responsible for electrolyte transport would be absent from the latter cells. Genetic and biochemical studies have since shown that this is not the case but that the LK state is determined by a single gene locus with two alleles, which are lacking in the HK type and which apparently are responsible for inhibition of cation transport and of the enzyme system. The genetically determined dimorphism may be due in part to a difference in the number of active sites per cell[39] or to the inhibition of the (Na–K) pump in low-K cells by intracellular potassium.[40] The pump in the HK cells also seemed to have a greater affinity for ouabain.[41] The M–L blood group antigen system in sheep[42] was associated with the HK–LK dimorphism insofar as the former had only M antigen and the latter L antigen. When antibodies were raised by injecting low-K cells into homozygous M sheep, these proved to have interesting properties. The inhibition of cation transport and of transport enzyme activity was reversed in a striking manner and a fourfold stimulation of these was realized by preincubating the LK cells or their ghosts with this anti-L serum.[43] As regards the mechanism of action, it was found that at 100% inhibition of ouabain-sensitive K influx, the anti-L serum treatment of LK cells increased ^{3}H-ouabain binding by their membranes.[39] This was interpreted as due to unmasking of pump sites, which would be expected to increase maximum transport rate V for the potassium ion uptake without necessarily modifying the affinity of the ions for the carrier.

The LK-type cells were found[38,44] to have higher passive fluxes as measured in the presence of ouabain than the HK type, which suggested that sites not used for active transport could become involved in this kind of transport. In support of this view, it appeared that K^+ uptake in LK cells showed saturation kinetics so typical of carrier-mediated transport. The lower passive fluxes in the HK cells on the other hand showed no sign of saturating with increase in $[K]_0$. The anti-L serum which increased active uptake of potassium in LK cells simultaneously decreased the passive fluxes of this cation. While these experiments certainly suggested an increased availability of pump sites under the influence of anti-L serum, equally strong evidence has been put forward for changes in the kinetics and in the relative affinity of pump sites for potassium and sodium within the HK and LK cells.

To examine the onset of the HK–LK dimorphism in sheep, which seemed to be absent until cell maturation, reticulocyte-rich fractions of red cells were obtained by gradient centrifugation of blood drawn from animals in which erythropoiesis had been stimulated by bleeding on three successive days (1 litre per day) and taking blood on the sixth day after the final bleeding.[45] In reticulocytes from the LK and HK types as well as in the mature cells of the latter type, the sodium pump was relatively insensitive to increase in $[K]_i$, whereas in mature cells of the former type the pump was inhibited by about 90% when $[K]_i$ was increased from 2 mM to 26 mM. In the HK cells, inhibition only amounted to 15% over the same concentration range and K uptake actually increased as $[K]_i$ was raised from 0 mM to 10 mM. Stimulation of the pump on increasing $[K]_i$ was first seen in red cells of goat[46] and later in human erythrocytes.[47] In normal human red cells, reduction of $[Na]_i$ reduced the rate of active transport of Na^+ and K^+, but only if intracellular potassium concentration was high.

When most of the cell potassium was replaced by choline or tetramethylammonium ions, the ouabain-sensitive excretion of sodium in exchange for potassium was increased even when $[Na]_i$ was below normal, and it was also inhibited by removal of external potassium.[48] When

[Na]$_i$ was maintained low and constant and [K]$_i$ varied, the activity of the sodium pump was an inverse function of the latter, so that intracellular potassium acted as a pump inhibitor under these conditions, but also at a normal level of cell sodium. This effect was independent of cell volume, [ATP]$_i$, and the methods used to alter intracellular cation concentrations. This was the inhibition seen in the HK cells above but was small in comparison with that found in the LK cells. The inhibition was apparently by competition with sodium ions for translocation sites on the internal side of the membrane.

In regard to the passive movement of potassium, it has been found[44] that intracellular potassium affected this movement in LK cells (where it is elevated) in the same manner in which it affected the active transport in HK cells, that is, stimulating it at low [K]$_i$ and moderately inhibiting it at higher [K]$_i$. All these results seem to indicate that the passive transport in the cell is mediated by inactive pump sites and that anti-L serum may act in part at least through the conversion of inactive pump sites into active forms.

A kinetic analysis[49] of the interaction of Na$^+$ and K$^+$ at internal sites of the membrane (Na$^+$ + K$^+$)-ATPase system of LK goat red cell "ghosts" seemed to confirm that potassium ions competed at each of the sodium-activating sites with a higher apparent affinity than sodium and that anti-L serum decreased K_m for sodium from 19.9 mM to about 3 mM, thereby decreasing the K/Na affinity ratio for the ouabain-sensitive ATPase system of the membranes. Contrary to expectations from previous findings, the maximum rate of hydrolysis of ATP did not appear to be affected by the anti-L serum.

CONFORMATIONAL CHANGES IN TRANSPORT ENZYME PRODUCED BY SODIUM AND POTASSIUM IONS

It seemed paradoxical that while in the LK cells potassium and rubidium ions were bound to internal sites with a higher affinity than sodium ions, they were not translocated. It should be stressed however that the ouabain-sensitive K efflux found in human red cells during partial or complete reversal of the pump does not use the "Na route."[31] One suggestion put forward for the paradoxical behavior is that the affinities for cation binding do not determine the subsequent fate of the cation–enzyme couples.[49] Another possibility was that the internal sites were in rapid turnover between two forms, E_a (enzyme) and E_b (*enzyme), with sodium having by far the highest affinity for E_a and cations such as potassium and rubidium being taken up preferentially by E_b. Even if sodium combined with a similar dissociation constant with both forms, the presence of potassium might displace the reaction toward formation of E_b by a magnitude which was depending also on the E_a/E_b equilibrium constant. If virtually all the E_a was in the bound form NaE_a, then efflux of K^+ by the pump machinery running forward would be prevented. Potassium and sodium would still compete for binding to E_b on the basis of the selectivity reported above. This is equivalent to saying that Na^+ but not K^+ can bring about a conformational change upon binding with the enzyme, and so "affinities" and "reactivities" of the cations become dissociated. The existence of a Na form and a K form of the dephosphorylated enzyme system has been well documented[50-53]; these forms bind ATP with different affinities, which determines whether ATP or P_i can phosphorylate. The latter reaction leads to reversal of the pump and, as we have seen, may be determined by the electrochemical gradient of these cations across the membrane and perhaps mediated by the internal presentation of the cation sites in question. The overall reactions may be summarized as follows[54]:

$$E_a + ATP \xrightleftharpoons{Mg^{2+}, Na^+} E_aP \cdot ADP \rightleftharpoons E_aP + ADP$$
$$\updownarrow Mg^{2+} \qquad (33)$$
$$E_bP \xrightarrow{K_+} E_b(K) + P_i \xrightarrow{ATP} E_a$$

where $E_b(K)$ is enzyme with occluded K, which is formed slowly from E_bP but rapidly when sufficient K^+ is present.

The kinetics of the enzyme processes in which ATP hydrolysis is coupled to cation transport have been examined[54] with purified $(Na^+ + K^+)$-ATPase from pig kidney and a fluorescent analog of ATP, formycin triphosphate (FTP), which undergoes a two- to threefold enhancement of this property on binding to the enzyme. The equilibrium binding of nucleotide to enzyme was therefore assessed from changes in the fluorescence signal. The FTP and also FDP were bound with a high affinity and could be displaced by excess of ATP or K^+. During turnover, at least with concentrations up to 24 μM FTP, the FDP was released before the rate-limiting step of the overall reaction, both in the presence and absence of potassium. When turnover was prevented by the absence of Mg^{2+}, the enzyme could still change its conformation depending on whether sodium or potassium was the predominant monovalent cation. But as pointed out above, the Na and K forms of the enzyme have different affinities for the nucleotide, so the rate of conformational change could be followed from the change in the fluorescence resulting from the release of FTP.

The changes in the affinity of the enzyme for FTP and FDP were used to examine the rates of interconversion of the two forms of the dephosphorylated enzyme: E_b, which had a low affinity for the nucleotide and which existed when potassium was the dominant cation; and E_a, with a high affinity for the nucleotides which existed when the sodium predominated. When the enzyme was preincubated with 0.8 mM K in the absence of sodium and magnesium and was then mixed with an equal volume of buffer containing 4 μM FTP and 100 mM Na, fluorescence increased, with a rate constant of about 0.25 sec^{-1}. Since the binding of FTP to the enzyme preincubated with sodium was very rapid (200 sec^{-1} at 4 μM FTP), it was considered likely that the rate-limiting step after preincubation with potassium was the interconversion of the K form to the Na form of the enzyme or some related process.

The conformational change in the reverse direction was studied with the enzyme which had been induced to bind FTP in the presence of 300 μM Na and then exposed suddenly to buffer containing 1 mM K. In this case FTP was replaced by FDP to avoid phosphorylation, and potassium was found to bring about a slow fall in fluorescence with a rate constant of 4.4 sec^{-1} (Fig. 5.2), compared with a fall of rate constant 110 sec^{-1} when the fluorescent nucleotide was displaced with excess of ATP. The slow change in fluorescence was once more interpreted as indicating the kinetics of a rate-limiting interconversion in this case from E_a to E_b. Although

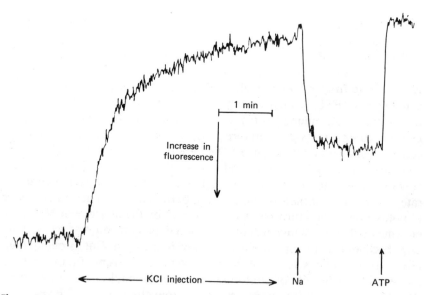

Figure 5.2 Dissociation of FDP from (Na$^+$ + K$^+$)-ATPase by addition of potassium ions. From S. J. D. Karlish, D. W. Yates & I. M. Glynn, *Nature* (Lond.) **263**, 251 (1976).

1 mM K should have been sufficient to displace virtually all of the FDP, its rate of displacement was still increased fivefold when K was increased from 1 mM to 5 mM although the total displacement was unaffected. This implied that the binding of potassium was to a low affinity site and that it accelerated the conformational change, possibly through the reactions

$$E_a + K^+ \rightleftharpoons E_a(K) \rightleftharpoons E_b(K) \tag{34}$$

where E_a is the form of the enzyme with a high affinity for nucleotide and a low affinity for K^+ and $E_b(K)$ is an enzyme with occluded potassium. Evidence for the existence of occlusion of K^+ in unphosphorylated ($Na^+ + K^+$)-ATPase has recently been found.[55]

POTASSIUM-ACTIVATED PHOSPHATASE AND THE SODIUM–POTASSIUM PUMP

The relationship between the ATPase activity and phosphatase activity of the red cell membrane has been examined[56] in cells incubated in solutions containing p-NPP, which seemed to enter the cells freely. Its intracellular concentration at equilibrium, which was reached with a half-time of 3 min was about a fourth of that outside, which was in agreement with the membrane potential. The active center of the K-activated membrane phosphatase is at the inner surface and therefore readily accessible to this substrate. When added to ATP-free red cells or reconstituted ghosts in the presence of sufficient Mg^{2+} to catalyze the phosphatase activity, p-NPP, contrary to previous reports,[57] did not seem to energize the transport of sodium or rubidium across the cell membrane.

Although it was recognized that P_i is a competitive inhibitor of membrane phosphatase activity and about 2 mM of this anion was present in the cells, it was considered unlikely that such inhibition was responsible for the failure of the substrate to support cation transport, since an increase of p-NPP well above the Michaelis constant for membrane phosphatase, at which point competitive inhibition by P_i should have been reversed, failed to bring about active cation transport. The maximum rate of K-dependent phosphatase activity[57] was about 0.6 m.mole·l^{-1} cells·hr^{-1}, and this must have been approached under the present conditions. This should have been sufficient to provide an extra 1.8 m.mole l^{-1} cells·hr^{-1} of ouabain-sensitive sodium loss and two thirds of this level of rubidium uptake if the hydrolysis of p-NPP were coupled to cation movement with the usual stoichiometry. No such coupling was evident.

When the concentration of p-NPP was increased progressively in the incubation medium of the ATP-containing cell, the active transport of Na and Rb was reduced to zero along a rectangular hyperbole. The half-maximum for this inhibition was 19 mM p-NPP. This substrate also blocked the Na–Na exchange to the same extent.

The inhibition here seemed to involve the combination of p-NPP with the active center of the membrane phosphatase at the inner surface of the membrane and was inversely related to [ATP]$_i$. It has been suggested[58] that phosphatase substrates phosphorylate the E_b conformation of the sodium pump to form E_bP, which then reacts with water in the presence of potassium to yield P_i. If the role played by E_bP in the overall pump activity is as shown in eq. (33) above,[59] then one would not expect either Na–K or Na–Na exchange to be driven by phosphorylation at this stage with formation of E_bP. If this phosphorylated enzyme is one of the stages through which the transport system passes cyclically during cation translocation, and if p-NPP can selectively react with E_b to yield E_bP, then we might expect the ATP-dependent Na–K and Na–Na exchanges to be progressively inhibited as the enzyme is increasingly trapped in the form E_bP with the raising of p-NPP concentration.

In addition to its effect on the ouabain-sensitive cation transport, p-NPP also increased ouabain-insensitive fluxes of both Na and Rb, the increase in passive permeability being four times greater in the case of Rb than with Na. This change was apparently mediated through a combination of p-NPP with sites at the inner surface of the membrane, half-saturating at 15 mM p-NPP, and was unaffected by the [ATP]$_i$. The increased ouabain-insensitive cation movement here at the expense of active transport is reminiscent of the greater passive permeability of LK-type red cells compared with the HK-type cells, where active transport was also found to be inhibited in the former cells.

ACTIVE TRANSPORT IN SKELETAL MUSCLES

Skeletal muscle has not proved to be the most convenient tissue in which to examine cation transport for a number of reasons. The apparent compartmentalization of ions into true intracellular space and into a space other than interstitital but containing high-Na and low-K concentrations[60-62] (possibly sarcoplasmic reticulum) has complicated the interpretation of sodium flux data by the introduction of a large element of exchange diffusion. Secondly, except for the case of the large muscle fibers of *Balanus nubilus*,[63] it has been difficult to change conveniently the internal composition of the fiber. Nevertheless the measurement of net transport of sodium and potassium ions in muscle has revealed a number of interesting and important properties of the pump there. For example, it was found[64] in the case of sodium-enriched frog sartorii that net excretion of this cation did not take place unless the energy gradient against which it had to be moved was less than a critical level of about 2 cal·meq^{-1} Na, referred to as the critical energy barrier for the excretion of sodium. The energy required was calculated from the equation

$$\frac{dG}{dn} = RT \ln \frac{[Na]_0}{[Na]_i} + E_m F \tag{35}$$

where dG/dn is the energy gradient per equivalent of Na; $[Na]_i$ and $[Na]_0$ are the concentrations of sodium in muscle fiber water and external medium, respectively; and E_m is the membrane potential measured by the microelectrode technique. It should be evident from this equation that the energy required for sodium transport may be reduced by lowering $[Na]_0$ or by reducing E_m through increase in $[K]_0$. Reduction of $[Cl]_0$ was also found to reduce E_m under these conditions, thereby promoting active transport. It was found that when $[Na]_i$ was about 59 m.mole·kg^{-1} fiber water and $[K]_0$ 10 mM, no net excretion of sodium took place when $[Na]_0$ was 120 mM, but when $[Na]_0$ was reduced to 104 mM, up to 20 m.mole Na·l^{-1} fiber water was extruded over a period of 30 min at room temperature, and an equivalent uptake of potassium took place.

Table 5.1 shows the quantities of sodium extruded at different calculated values of energy requirement under various conditions. The net

Table 5.1 Energy Barrier to Sodium Excretion by Sodium-Rich Sartorii Immediately After Reimmersion[a]

Procedure	Membrane potential E_m (mV)	$E_m F \cdot RT \cdot \ln [Na_o]/[Na_i]$ (cal·meq^{-1} Na)		Total (cal·meq^{-1} Na)	Average Na excretion (meq·kg^{-1})
120,0/120,10	67.3 ± 1.5	1.55	0.44	1.99	1.4 ± 2.4
120,0/115,10	(66.0)	1.52	0.41	1.93	3.7 ± 2.3
120,0/110,10	(64.7)	1.49	0.38	1.87	18.2 ± 1.7
120,0/104,10	63.4 ± 0.7	1.46	0.35	1.81	19.6 ± 1.5
120,0/80,10	56.8 ± 0.8	1.32	0.22	1.54	19.8 ± 1.8
120,0/120,30	48.9 ± 0.7	1.12	0.44	1.56	14.6 ± 1.8
120,0/120,60	41.6 ± 1.0	0.96	0.44	1.40	15.6 ± 2.9
104,0/104,10	71.7 ± 0.9	1.65	0.51	2.16	1.0 ± 0.9
104,0/80,10	54.8 ± 0.7	1.27	0.39	1.66	12.6 ± 2.2
104,0/104,30	46.6 ± 2.2	1.07	0.51	1.58	9.0 ± 1.7

[a] Parentheses under membrane potentials indicate interpolated values.[64]

transport here was inhibited by addition of ouabain or iodoacetate to the recovery fluid. On the other hand, the addition of insulin or of isoproterenol greatly stimulated the sodium pump. For example,[65] when 100 U·l^{-1} of insulin was added to recovery fluid containing Na-rich sartorii, sodium extrusion took place even in the presence of 120 mM Na and 2.5 mM K where the calculated energy gradient was as high as 3.7 cal·meq^{-1} sodium extruded, which was nearly twice the critical energy barrier in the absence of this hormone. The hormone therefore appeared to increase the energy available to the sodium pump.

A 50% stimulation of ^{24}Na efflux from rat diaphragm on addition of insulin was subsequently confirmed[66] with reduction of the final steady-state concentration of this cation in the fibers. An increased accumulation of potassium in muscles which had been noted previously[67,68] on addition of insulin to frog muscles bathed with normal Ringer's fluid and was usually associated with an increased oxygen consumption by the tissue was probably the result of membrane hyperpolarization which occurred in the presence of this hormone.[69] Before dealing with the possible mechanism of action of insulin in relation to the cation pump, it might be profitable to consider the origin of the membrane hyperpolarization.

ELECTROGENIC SODIUM PUMP AND POTASSIUM UPTAKE

In spite of some indications to the contrary,[70,71] it was generally accepted in the mid-1950s when Hodgkin delivered his Croonian Lecture[72] to the Royal Society that the Na–K pump in nerve and muscle was electrically neutral, transporting one potassium ion into the cell for each sodium ion extruded during a cycle of its operation. Indeed attempts to detect a contribution by the sodium pump in muscle to membrane potential[73-75] had proved negative. In one case,[75] although sodium enrichment of fibers by presoaking of isolated frog sartorii in cold potassium-free fluid had been carried out in order to increase the level of sodium transport, the absence of calcium ions from the recovery fluid may have accounted for the apparent absence of electrogenic transport of sodium.

When muscles enriched with sodium ions as described above were reimmersed in Boyle–Conway Ringer containing 10 mM K and 104 mM Na and membrane potential was measured and compared with calculated equilibrium potential for potassium based on analysis of the muscles, the former potentials were found to be more negative than the latter by about 11 mV when measured about 10 min after immersion in recovery fluid.[76] The difference between E_m and E_K was increased considerably when external sodium was replaced by choline chloride to facilitate sodium excretion[77] or when the experiment was carried out in fluids in which chloride was replaced by the nonpermeable anion methylsulfate to reduce the short-circuiting effects of anion movement.[78] The electrogenic pumping of sodium from skeletal muscles[79-82] and from nerve[83,84] has been confirmed and a number of reviews have been written on electrogenic pumping in these and other tissues.[85-88] The ability of the Na–K pump in skeletal muscle to hyperpolarize the membrane[85-88] by extruding sodium at a faster rate than that of potassium uptake, coupled with the relatively high membrane permeability to the latter cation, has raised the question of the extent to which potassium accumulation is passive and whether it takes place through permeability channels, as opposed to active carriage via the pump.

The ability of barium ions to block potassium channels has been exploited[89] in an attempt to answer this question. In the presence of 5 mM Ba^{2+}, the influx of ^{42}K into sodium-rich muscles in recovery fluid gave a good approximation of the total potassium accumulation and was completely abolished by addition of ouabain. Practically all the passive ^{42}K influx was barium sensitive and was decreased as $[Na]_i$ was increased. The calculated active uptake of potassium in the presence of barium was 25 $\mu mole \cdot g^{-1} \cdot hr^{-1}$, and the calculated passive influx was 1.25 $\mu mole \cdot g^{-1} \cdot hr^{-1}$, so that over 90% of net uptake appeared to be through the pump. Kinetic studies of K^+ influx in the presence and absence of barium and/or ouabain indicated however that the addition of barium induced the pump to transport more potassium to compensate for the fall in P_K.

When membrane potentials were compared during sodium transport[79] in recovery fluid containing either 10 mM K or Rb (which has about one third the permeability of K), the membrane was hyperpolarized to a greater extent in the latter solution, and it was calculated that not more than about 10% of the Rb^+ entered the fibers passively, the remainder being exchanged from sodium ions on the carrier molecule. The antihistamine drug mepyramine and a number of local anesthetics have been found to increase the membrane resistance of skeletal muscle, apparently through decrease in permeability to ions including potassium. When added[80] to sodium-enriched muscles pumping sodium into recovery fluid in the presence of 5 mM K, they greatly increased the membrane potential, indicating the degree to which passive movement of potassium contributed to the short circuiting of the potential generated by the pump.

EFFECT OF INNERVATION ON POTASSIUM TRANSPORT AND PERMEABILITY

The isolation of skeletal muscles with about 1 inch of nerve attached has been found to make the pump less electrogenic. When rat extensor digitorum longus (EDL) muscles were dissected out of the animal with nerve still attached and soaked for 2 hr at about 2°C in K-free Krebs fluid with 160 mM Na and one muscle of each companion set separated from the nerve before both were reimmersed at 37° in recovery fluid with 10 mM K and 137 mM Na, it was found[90] that the innervated muscles transported almost twice as much sodium and potassium as their isolated companion and had membrane potentials that were not significantly different from E_K. The same behavior in the presence of innervation was found in the case of frog muscles.[91] These results have suggested that the isolation of muscle from nerve may have increased membrane resistance through a decrease in P_K or changed the coupling ratio on the pump, thereby hyperpolarizing the membrane during active transport of ions.

When the influx of ^{42}K was measured[92] in rat muscles *in vivo* within 48 hr of unilateral section of the sciatic nerve, it was found to decrease in response to denervation under conditions where $[K]_i$ was actually increasing. It was concluded that nerve section decreased the unidirectional fluxes of potassium but reduced efflux to a greater extent than influx. Under these conditions E_m has been found to decrease,[93-95] which would be in keeping with a decrease in the P_K/P_{Na} ratio. Under the circumstances and in the absence of strong evidence for a change in the coupling ratio of the pump, it seems likely that the effect of denervation on E_m and net transport from Na-rich muscles may be explained by a decrease in P_K. It should be noted however that in the case of denervation of four to six days' duration, the undirectional ^{42}K fluxes appeared[92] to increase with a net loss of potassium from the fibers in exchange for sodium. An increase in g_K accompanied by a fall in g_{Cl} has been found in rat skeletal muscles two weeks after denervation.[96]

STIMULATION OF THE SODIUM PUMP BY INSULIN AND BY β-ADRENERGIC AGENTS

It has been noted[65] that the membrane hyperpolarization induced by insulin[97] may be due to stimulation of the electrogenic sodium pump. However, while the evidence for this effect is strong, there is another aspect of insulin action on the membrane which should not be overlooked, namely, its apparent reduction of the unidirectional fluxes of potassium in freshly dissected skeletal muscle of rat,[98] an effect which was more pronounced in hypophysectomized animals. While these fluxes were reduced, there was still a net loss of sodium and net gain of potassium in the muscle; so it seems that K efflux was reduced more than K influx.

As to the mechanism by which insulin might stimulate transport of sodium and potassium, the following possibilities have been considered: (a) increase in the energy-rich phosphate in the muscle fibers[99]; (b) increase in active pump sites in the membrane[100]; and (c) increase in the affinity of existing sites for sodium over other cations.[101] The specific binding of ^{3}H-ouabain to frog muscle fibers was increased in the presence of insulin, and this binding correlated with pump inhibition. It was therefore assumed that more binding sites appeared on the membrane surface. As this occurred even in the presence of cycloheximide, which blocks protein synthesis, it was concluded that the hormone acted by unmasking existing sites rather than promoting synthesis of new carrier.

Insulin has been found[102] to stimulate membrane-bound $(Na^+ + K^+)$-ATPase prepared from frog skeletal muscles, and this effect was more pronounced when the concentration of sodium and ATP were reduced in the medium, indicating that k_m rather than V for sodium was affected. Stimulation increased from 10–53% in the presence of 25 mM Na and 2 mM ATP to 400% in 4 mM Na and 0.5 mM ATP both in the presence of 10 mM K. At 100 mM Na, 2 mM ATP and 20 mM K, where the ATPase activity was maximal, the hormone was without effect. The stimulation of the ATPase seemed to involve an increasing affinity to ATP as well as to sodium, since with 100 mM Na present, insulin stimulated the ATPase only when [ATP] was reduced.

Catecholamines also stimulated the Na–K pump in muscle, including frog sartorius,[103] rat diaphragm,[104] and soleus.[105] The importance of the β action of these substances was evident in the fact that addition of 5×10^{-8} g·ml^{-1} isoprenaline[90] increased the net transport of ions from Na-rich soleus muscles of rat by 50%, producing also membrane hyperpolarization[106] which was blocked by propranolol[105] and by cooling of the tissue to 18°C. Adrenaline also produced hyperpolarization in avian slow muscles[107] that was ouabain sensitive, and this effect was believed to be due to the electrogenic pumping of sodium. The effect seemed to be more pronounced in red soleus muscle than in the paler extensor digitorum longus (EDL) muscles, and this corresponded with the greater responsiveness of the sarcolemmal adenyl cyclase system in the former muscles[108] to stimulation by isoprenaline.

In rat soleus muscle, $6 \times 10^{-6} M$ adrenaline[109] increased ^{22}Na efflux by 83% and ^{42}K influx by 34%, while membrane potential was increased by 10%. The steady-state ratio $[K]_i/[Na]_i$ was also increased under the influence of adrenaline. Stimulation of ^{22}Na efflux was potentiated by $2 \times 10^{-3} M$ theophylline when adrenaline concentration was low in the recovery fluid. It seems likely that the adenyl cyclase and cyclic AMP systems were mediating the hormone effect in this case,[109,110] but calcium ions may also have been involved.[104]

β-Adrenergic receptors have been detected[111] in transverse tubules from rabbit skeletal muscle with a density of 0.61 p.mole·mg^{-1} protein but were evidently absent from the terminal cisternae and longitudinal reticulum of the sarcotubular system. In the microsome fraction prepared from tubular membranes, isoprenaline stimulated adenyl cyclase. Such stimulation has been found[112-114] to promote calcium accumulation in these particles. The stimulation of sodium transport observed in muscle in the presence of β-adrenergic agents may be mediated through exchange of sodium and calcium across the membrane.

It was also found that catecholamines enhanced (Na$^+$ + K$^+$)-ATPase activity in membrane fragments prepared from rat skeletal muscles[115] and from cardiac tissue.[116] Although isoprenaline appeared to increase lactate production, glycogen breakdown, and the utilization of alanine, perhaps in gluconeogenesis, it had no effect on total ATP or creatine phosphate levels of the cell.[117]

POTASSIUM AND SODIUM TRANSPORT IN CARDIAC TISSUE

Cardiac papillary muscle has been used[118] in the study of active transport of sodium and potassium. When this was cooled to 2–3°C in physiological saline to inhibit the pump, [K]$_i$ decreased from 160 m.mole·kg^{-1} to 120 m.mole·kg^{-1} fiber water and [Na]$_i$ increased to 80 m.mole·kg^{-1} fiber water after about 90 min. On rewarming the muscle to 30°C for 1 hr in the same solution, [K]$_i$ increased to 170 m.mole·kg^{-1} while [Na]$_i$ was reduced to 45 m.mole·kg^{-1} fiber water due to active transport. Recovery did not take place unless [K]$_0$ was greater than 1 mM. Although the pump was blocked by addition of 10^{-6} M ouabain, this was reversed by increasing [K]$_0$ to 25 mM, suggesting that potassium and ouabain competed for the same carrier sites.

A connection may have been established between the inotropic action of the cardiac glycosides and their inhibition of the Na–K pump in that muscle.[119] It has been suggested,[120] for example, that increased influx of sodium and the resulting increased intracellular concentration of this cation might mediate the increase in muscle contractility. This suggestion has been supported by the finding that conditions which lead to increase in [Na]$_i$ also have an inotropic effect on the muscle. For example, removal of external potassium, the blocking of the pump with p-chloromercuribenzoate, NEM, or ethacrynic acid[121] or the addition of sodium ionophores to the bathing fluid all exerted an inotropic action on the heart. Partial replacement of external sodium by choline reduced the rate of development of the inotropic effect of ouabain.[122] Cardiac glycosides and their aglycones bind to the $(Na^+ + K^+)$-ATPase on the outside of the cell at the stage in the enzyme cycle when the transport of sodium has been completed in the presence of Na^+, Mg^{2+}, and ATP. However the combination of potassium with the phosphorylated enzyme converts it to a form which seems to be less accessible to the drug.[126] Therefore, while increase in $[Na]_i$ should favor binding of the cardiac glycoside by increasing transfer of carrier sites to the outside of the membrane, increase in $[K]_0$ should prevent its combination. As a result, when plasma potassium concentration was increased in the case of digitalis-treated dogs, a larger dose was needed to produce a comparable inotropic effect. In addition to inhibiting the binding of the glycosides to $(Na^+ + K^+)$-ATPase, potassium ions also delayed their release from the enzyme *in vitro*, so there was a delayed loss of inotropic response when potassium was increased in drug-free fluid used to perfuse isolated hearts.

Although the concentration of glycoside which has been used to evoke the positive inotropic effect ($10^{-9}\,M$) was several orders of magnitude less than that used to inhibit the Na–K pump *in vitro* ($10^{-6}\,M$), high affinity binding sites for ouabain ($k_i = 10^{-8}$) have also been reported[123] in some species which have a high sensitivity to the drug. The aglycone strophanthidin has been found[124] to increase the slow inward current in the Purkinje fibers of calf hearts while bringing about a twofold increase in twitch tension within 3 min of exposure. During longer exposure the cardiotoxic effects of these substances become evident in a loss of cell potassium, detectible with K-selective microelectrodes,[125] as might be expected during pump inhibition.

POTASSIUM TRANSPORT IN SMOOTH MUSCLE

In the case of smooth muscle as exemplified by guinea pig taenia coli,[126] when this was sodium enriched and placed in recovery fluid in which sodium had been replaced by magnesium ions and ^{42}K was added, its uptake depended on $[K]_0$ in a sigmoid manner with a half-maximum for activation of 4 mM. Its maximum rate of uptake was 3 m.mole·kg^{-1}·min^{-1}. When sodium was present in the external fluid, a greater $[K]_0$ was required to obtain the same degree of ^{42}K uptake, and this was interpreted as showing competition between these cations for external carrier sites. A ouabain-sensitive ^{24}Na efflux into Na-free solution required $[K]_0$ of 1–2 mM for half-maximum activation. When tritiated ouabain was used[127] to measure Na–K pump sites in smooth muscle of the taenia coli, two phases of binding were noted (Fig. 5.3). The first, saturating at low ouabain concentration and sensitive to changes in $[K]_0$,

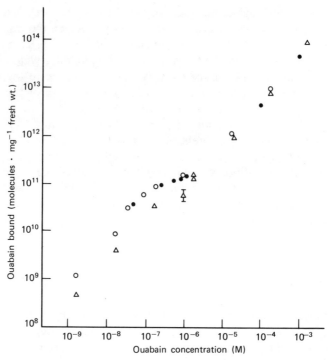

Figure 5.3 Binding of ³H-ouabain to tissue at different glycoside concentrations, and in Krebs solution containing zero K (open circles), 0.6 mM K (closed circles), and 6 mM K (open triangles). Tissue was pretreated for 1 hr at 36°C in Krebs fluid with desired [K]$_0$ before exposure of the guinea pig taenia coli to a similar solution containing ³H-ouabain for a further 1 hr. From A. F. Brading & J. H. Widdicombe, *J. Physiol.* **238**, 235 (1974).

accounted for 1.1×10^{11} molecules·mg^{-1} tissue or about 250–300 molecules·μm^{-2} (assuming volume-to-surface ratio of 1.5). The other, saturating at higher concentrations, was probably nonspecific. Only 60% of ⁴²K uptake was blocked however by the maximum ouabain concentration used in normal Krebs fluid.

Before leaving smooth muscle, the possible physiological significance of electrogenic pumping of sodium in vascular muscle should be mentioned. Here it was found that vasodilatation produced by potassium ions *in vivo* following denervation or adrenergic blockade was not consistent with a change of potassium diffusion potential. When perfusion pressure at constant blood flow was plotted against change in plasma potassium concentration,[128,129] a graded decrease in pressure was produced as [K]$_0$ was increased up to twice its normal value. On the other hand decrease in plasma potassium down to about 10% of normal produced an almost linear increase in perfusion pressure. Decrease in pressure correlated with the hyperpolarizing response to increase in $[K]_0$ over this range of concentration, suggesting that the steady-state potential might be influenced directly by an electrogenic sodium pump.[130]

SODIUM AND POTASSIUM TRANSPORT IN DIALYZED SQUID AXONS

While perfused axons have been used successfully in the study of the origin of the action potential, they have been less satisfactory in the study of active transport. Internal dialysis of axons,[131,132] which has largely overcome this difficulty, involves the insertion of a glass tube 250 mm long and 0.175 mm in diameter into the axon (Fig. 5.4). A central 20 mm length of this tube was porous with about 10% of the surface having pores 20–40 Å in diameter, which allowed ions and small molecules such as ATP through but were virtually impermeable to proteins. When solutions of the required composition were passed through this capillary at the rate of 0.6–1.2 $\mu l \cdot min^{-1}$, equilibrium was reached with the axoplasm with a time constant of 3–5 min. When ^{22}Na was added to the dialyzate and ^{24}Na and ^{42}K to the external bathing fluid, unidirectional fluxes could be measured under a variety of conditions. For example, when $[ATP]_i$ was

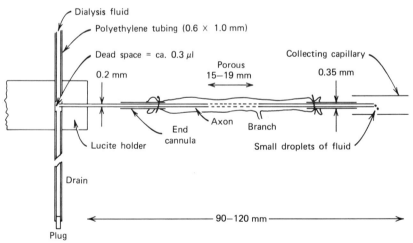

Figure 5.4 Diagram (not to scale) of lucite holder for the porous capillary, which is mounted on a micromanipulator on left. In the center, the capillary is shown inserted into a squid axon; and on the right, the capillary which collects the effluent fluid being pumped through by a motor driven syringe. From F. J. Brinley & L. J. Mullins, *J. Gen. Physiol.* **50**, 2303 (1967).

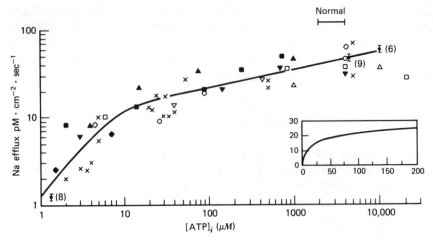

Figure 5.5 Relationship between $[ATP]_i$ and the Na efflux in 47 dialyzed squid axons is shown on a log–log plot. A horizontal bar indicates the range of ATP in axoplasm extruded from freshly dissected axons taken from living squid. Insert shows data for range 0–200 μM ATP plotted on a linear scale. Data normalized to 15°C. From F. J. Brinley & L. J. Mullins, *J. Gen. Physiol.* **52**, 181 (1968).

reduced from its normal value of 4.4 mM to about 1 μM through dialysis with ATP-free dialyzate, ^{22}Na efflux fell from its control value of 40 p.mole·cm^{-2}·sec^{-1} down to 1.3 p.mole·cm^{-2}·sec^{-1}. Efflux of ^{22}Na showed a nonlinear response to increase in $[ATP]_i$ (Fig. 5.5), rising more steeply in the range of 1–10 μM than over the range of 10–10,000 μM ATP.[133] When $[ATP]_i$ was maintained constant at its normal level and $[Na]_i$ varied, ^{22}Na efflux was linearly related to the latter over the range of 5–230 mM $[Na]_i$.

The potassium response of Na efflux and also its strophanthidin sensitivity were determined by $[ATP]_i$ or more correctly by the $[ATP]_i/[ADP]_i$ ratio. The Na efflux was found to be maximally dependent on $[K]_0$ when this ratio was 10:1; it was insensitive when the ratio was 1:2; and efflux was even inhibited by addition of $[K]_0$ when this ratio was 1:10.[134] At best the dependence of sodium efflux on external potassium concentration was much less than might have been expected. When squid axons were immersed in K-free artificial seawater, for example, ^{22}Na efflux fell to 30–50% of controls, compared with 3% in the virtual absence of ATP from the axons. Likewise the influx of potassium into axons containing minimal concentration of sodium and ATP was about 6.0 ± 0.0 p.mole·cm^{-2}·sec^{-1} increasing to 19.0 ± 0.07 p.mole·cm^{-2}·sec^{-1} when these were at their normal values. When this change was compared with

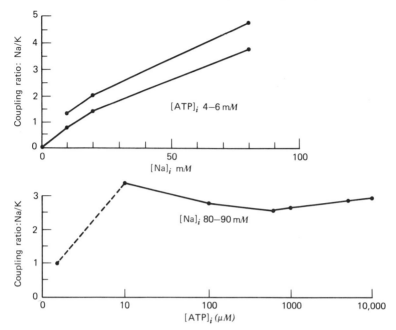

Figure 5.6 Upper graph shows dependence of the apparent coupling ratio Na/K on $[Na]_i$, for (upper curve) an assumption that K influx under internal Na-free conditions is a baseline from which coupling is to be computed. The lower curve shows the apparent coupling ratio between the ATP-dependent K influx and Na efflux. Lower graph shows Na/K coupling ratio as a function of $[ATP]_i$. Values below 10 μM ATP are subject to large errors in flux estimates. From F. J. Brinley & L. J. Mullins, *Ann. N.Y. Acad. Sci.* **242**, 406 (1974).

the 31-fold attenuation of Na efflux on reducing $[ATP]_i$ to a minimum, it was obvious that transport of these ions in squid axons was not comparable with that found in erythrocytes, and it led Mullins to speculate on the existence of independent sodium and potassium pumps in this tissue.[135]

This independence was also indicated by the rather variable coupling ratio of the sodium–potassium exchange measured with ^{22}Na and ^{42}K where $[ATP]_i$ was maintained at its normal level and $[Na]_i$ increased over the range of 10–100 mM. Here the ratio (Fig. 5.6) increased linearly up to a maximum of 4:1 with increase in $[Na]_i$. When $[Na]_i$ was maintained at 80 mM and $[ATP]_i$ varied, the coupling ratio was 3:1 over the range of 10–1000 μM ATP. Finally the effect of the external potassium concentration on the coupling ratio was examined. When $[K]_o$ was low, much more Na was extruded than potassium taken up; but at 10–20 mM K there was little further increase in Na efflux, while the influx of K continued to increase in proportion to $[K]_o$. However the limiting ratio for saturation with external potassium was still greater than 3:1. Such high coupling ratios for Na:K exchange strongly suggest electrogenic pumping of sodium in nerve.

POTASSIUM TRANSPORT IN LIVER

Intracellular concentrations of ions in dog liver based on extracellular volume measured by tritiated inulin were found to be as follows: $[K]_i$, 172 ± 13; $[Na]_i$, 22.1 ± 4.4; and $[Cl]_i$, 24.1 ± 2.9 m.mole·l^{-1} cell water. The diffusion potential based on the Goldman equation, where P_{Na}/P_K was taken to be 0.17, was found to be −37 mV, compared with −44.4 mV measured by microelectrodes.[136] Within minutes of adding $10^{-4} M$ ouabain to the medium, E_m became less negative, suggesting inhibition of an electrogenic sodium pump, as also indicated by the Na:K coupling ratio of 3:2.[137] On the other hand, partial depletion of ATP by hypoxia or by addition of $10^{-2} M$ antimycin or $2 \times 10^{-2} M$ fructose produced membrane hyperpolarization accompanied by potassium loss from the liver. In rat liver cells,[138] $10^{-3} M$ cyanide reversibly depolarized the membrane by 35%, while ouabain even at a concentration of $10^{-2} M$ had no effect on membrane potential. The ouabain inhibited Na efflux and K influx by 58% and 72%, respectively.

POTASSIUM UPTAKE IN YEAST AND BACTERIA

When Baker's yeast was fermenting glucose in the presence of $0.2 M$ KCl, there was a 1:1 exchange of H$^+$ excreted for K$^+$ taken up, with the latter reaching a concentration of $0.6 M$ within the cell.[139] This uptake of K$^+$ was probably active in view of the low membrane permeability of this cation and the fact that a membrane potential of about −137 mV would be required to account for its passive uptake. In the resting state, $[K]_i$ was about 50 times greater than $[Na]_i$, indicating a greater affinity of potassium for the carrier sites. The measurement of ion fluxes and the application of the Lineweaver–Burk double-reciprocal plots to the resulting data[140] have given relative affinities for the carrier of 1380, 100, 3.8, 4.2, and 0.5 for H$^+$, K$^+$, Na$^+$, Rb$^+$, and Mg$^+$, respectively. While competitive inhibition seemed to take place between potassium and rubidium at the K carrier sites, the interaction was more complex in the case of other cations. For example, at low pH the selectivity of the carrier for K over Na seemed to increase.[141]

The selective displacement of ^{42}K from external membrane sites by rubidium or by magnesium at 4–5°C has been used[142] to measure the capacity of K carrier sites on the cell surface, which was found to be 0.1–0.2 m$M \cdot$kg^{-1} of yeast. The active transport of protons and potassium ions in yeast has been blocked[143] to varying degrees by a number of metabolic inhibitors including DNP, azide, and iodoacetate and also by cyanide in the absence of both glucose and phosphate. Although these may have acted by preventing the synthesis of ATP, cation transport was not ouabain sensitive in yeast. Conway's theory of a redox pump using electron energy and the respiratory chain directly to release of protons at the outer surface of the cell membrane, a theory now much in vogue in the case of mitochondrial hydrogen ion pumping, still seems to be applicable to yeast. It was suggested[144] that protons were released into the external medium during the transfer of electrons from a hydrogen carrier such as a flavoprotein to a cytochrome, which accepted only the electrons. The electrons then returned to the inner surface of the membrane where they produced an equivalent alkalinization.

The marked effect of redox dyes on transport of hydrogen and potassium ions in fermenting yeast[145] has been put forward as evidence for the participation of the respiratory chain in the process. The mean redox potential of fermenting yeast measured in pure nitrogen by means of a bright platinum electrode was about 180 mV. Reducing dyes ($E_h < 180$ mV) inhibited transport of H$^+$ and K$^+$ by the fermenting yeast while oxidizing dyes ($E_h > 180$ mV) increased their transport above that of controls. Although it was suggested that the reducing dyes might have shunted hydrogen atoms at the critical region of the respiratory chain, thereby preventing proton release or, in the case of oxidizing dyes, stimulate reduction of the final acceptor, they might also have acted on glutathione in the membrane. Oxidation of the latter by methylene blue has been found to be associated with increased extrusion of protons into the medium.[146]

A respiratory-deficient mutant of yeast in which the synthesis of cytochromes was inhibited[147] also behaved quite differently from normal yeast in transporting ions. For example, oxidizing dyes did not increase net transport of hydrogen and potassium ions[148]; and in mitochondria prepared from such yeast no ATP-dependent K$^+$ uptake could be demonstrated.[149] So it seems possible that the terminal segment of the respiratory chain no longer functions in these cells.

Phosphate uptake seemed to be dependent on potassium accumulation in yeast. Its content in yeast was about 120 m.mole$\cdot$kg^{-1} wet weight of which about a tenth was present as orthophosphate. The membrane was relatively impermeable to this anion, but in the presence of glucose it entered as H$_2$PO$_4^-$, its uptake being limited by the internal acidity which it produced. The k_m for its uptake was $4 \times 10^{-4}\,M$. Phosphate influx in turn facilitated the accumulation of magnesium ions. When yeast had been made potassium rich by fermentation in the presence of potassium chloride,[150] it subsequently accumulated significantly more phosphate than yeast not so pretreated, but this was probably due to a symport of protons and phosphate rather than to antiport against potassium.

In bacteria, also phosphate uptake was potassium dependent, which also limited growth and metabolism.[151] However in this case preloading of the cells with potassium did not stimulate uptake of the anion on subsequent exposure to phosphate, so here a symport with potassium rather than with protons as in yeast was indicated. When the electrochemical gradient for protons was large in the case of yeast (pH_0 4.5 at 30°C) and the cell had been ATP depleted by inhibition of metabolism, a spontaneous release of potassium and equivalent proton uptake was observed,[152] which was accelerated in the presence of proton conductors such as DNP. The addition of glycine to the external medium also facilitated this downhill movement of ions, and the amino acid was taken up in the process. This seems like an interesting case of nonelectrolyte transport at the expense of electrolyte movements, especially as glycine uptake from a 5 μM solution was 400 times faster than at pH 7.4 when the external pH was reduced by about three units.

The role of potassium in the maintenance of osmotic pressure in bacterial cells has been emphasized through experiments[153] in which its uptake was triggered by a sudden increase in the osmolarity of the medium (an osmotic shock) not involving increase of external potassium concentration. It seems likely that there exists a mechanism which regulates K influx (at constant efflux rate) very precisely in response to variation of external osmotic pressure. It had been noted that K-depleted cells had high K uptake rates, k_m for this process being in the millimolar range, while K–K exchange in the steady state occurred at a much lower rate and with a very high affinity for potassium.[154]

The kinetics of K–K exchange in the steady state and of net uptake following osmotic shock were studied in four genetically distinguishable K transport systems in *Escherichia coli,* particularly in relation to energy sources. In one such system, namely, TrkA with k_m of 1.5 mM and V of 55 μ.mole·g^{-1}·min^{-1}, both ATP and a protonmotive force were required in order to carry out net uptake of K^+, a situation apparently unique for bacterial transport. On the other hand, K–K exchange within this system required only ATP. It seems that the cells must sense turgor pressure and respond by regulating potassium transport. As all systems examined had maximum rates of uptake exceeding that required for growth (6 μ.mole·g^{-1}·min^{-1} with doubling time of 60 min), all systems must have been regulated by osmotic feedback. Once cell K had risen to a new equilibrium dictated by the higher osmolarity of the medium, influx fell to the control value and net influx ceased. The protonmotive force was usually constant and in close equilibrium with the energy of ATP in bacteria.[155] It was suggested that the TrkA system might be regulated by the electrical component of the protonmotive force (see Chapter 6). An increase in turgor pressure might result in smaller ΔE and larger ΔpH.

REFERENCES

1. L. Lambotte, *J. Physiol.* **269**, 53 (1977).
2. U. V. Lassen & O. Sten-Knudsen, *J. Physiol.* **195**, 681 (1968).
3. G. Gardos, *Acta Physiol. Acad. Sci. Hung.* **6**, 191 (1954).
4. B. E. Hailey & J. F. Hoffman, *Proc. Natl. Acad. Sci. U.S.* **71**, 3367 (1974).
5. S. J. D. Karlish, P. J. Jørgensen & C. Gitler, *Nature* (Lond.) **269**, 715, (1977).
6. P. L. Jørgensen, *Biochim. Biophys. Acta* **356**, 36 (1974).
7. J. Kyte, *J. Biol. Chem.* **247**, 7642 (1972).
8. P. L. Jørgensen, *Biochim. Biophys. Acta* **356**, 53 (1974).
9. A. Klip & C. Gitler, *Biochem. Biophys. Res. Commun.* **60**, 1155 (1974).
10. K. Sigrist-Nelson, H. Sigrist, T. Bercovici & C. Gitler, *Biochim. Biophys. Acta* **468**, 163 (1977).
11. P. L. Jørgensen, *Biochim. Biophys. Acta* **401**, 399 (1975).
12. P. L. Jørgensen, *Biochim. Biophys. Acta* **446**, 97 (1977).
13. J. Jensen & P. Ottolenghi, *Biochem. J.* **249**, 5234 (1976).
14. H. J. Schatzman, *Nature* (Lond.) **196**, 677 (1962).
15. K. P. Wheeler & R. Whittam, *J. Physiol.* **207**, 303 (1970).
16. S. S. Goldman & R. W. Albers, *J. Biol. Chem.* **248**, 867 (1973).
17. D. E. Richards, P. J. Garrahan & A. F. Rega, *J. Membr. Biol.* **35**, 137 (1977).
18. I. M. Glynn, J. F. Hoffman & V. L. Lew, *Phil. Trans. Roy. Soc. B*, **262**, 91 (1971).
19. J. C. Skou, *Biochim. Biophys. Acta* **23**, 394 (1957).
20. J. C. Skou, *Biochim. Biophys. Acta* **42**, 6 (1960).
21. I. M. Glynn, *J. Physiol.* **160**, 18P (1962).
22. R. Whittam, *Biochem. J.* **83**, 29P (1962).
23. R. W. Albers, *Ann. Rev. Biochem.* **36**, 727 (1967).
24. I. M. Glynn & S. J. D. Karlish, *Biochem. Soc. Spec. Publ. 4*, 145 (1974).
25. P. J. Garrahan & I. M. Glynn, *J. Physiol. 192*, 237 (1967).
26. P. J. Garrahan & I. M. Glynn, *J. Physiol. 192*, 159 (1967).
27. P. J. Garrahan & I. M. Glynn, *J. Physiol. 192*, 189 (1967).

28. P. J. Garrahan & I. M. Glynn, *J. Physiol.* **192**, 217 (1967).
29. R. Blostein, *J. Biol. Chem.* **245**, 270 (1970).
30. I. M. Glynn, S. J. D. Karlish, J. D. Cavieres, J. C. Ellory, V. L. Lew & P. L. Jorgensen, *Ann. N.Y. Acad. Sci.* **242**, 357 (1974).
31. I. M. Glynn, V. L. Lew & U. Lüthe, *J. Physiol.* **207**, 371 (1970).
32. T. J. B. Simons, *J. Physiol.* **237**, 123 (1974).
33. P. L. Jørgensen, *Q. Rev. Biophys.* **7**, 239 (1975).
34. R. Whittam, C. Hallan & D. G. Wotton, *Proc. Roy. Soc. Lond.* **B193**, 217 (1976).
35. E. Abderholden, *Hoppe-Seyl. Z.* **25**, 65 (1898).
36. J. V. Evans, *Nature* (Lond.) **174**, 931 (1954).
37. J. V. Evans & A. T. Phillipson, *J. Physiol.* **139**, 87 (1957).
38. D. C. Tosteson & P. G. Hoffman, *J. Gen. Physiol.* **44**, 169 (1960).
39. P. B. Dunham & J. F. Hoffman, *J. Gen. Physiol.* **58**, 94 (1971).
40. P. G. Hoffman & D. C. Tosteson, *J. Gen. Physiol.* **58**, 438 (1971).
41. C. H. Joiner & P. J. Lauf, *J. Membr. Biol.* **21**, 99 (1975).
42. B. A. Rasmussen & J. G. Hall, *Science* (N.Y.) **151**, 1551 (1966).
43. J. C. Ellory & E. M. Tucker, *Nature* (Lond.) **222**, 477 (1969).
44. P. B. Dunham, *J. Gen. Physiol.* **68**, 567 (1976).
45. P. B. Dunham & R. Blostein, *Biochim. Biophys. Acta* **455**, 749 (1976).
46. J. R. Sachs, J. C. Ellory, D. L. Kropp, P. D. Dunham & J. F. Hoffman, *J. Gen. Physiol.* **63**, 389 (1974).
47. R. P. Garay & P. J. Garrahan, *J. Physiol.* **231**, 297 (1973).
48. A. B. Knight & L. G. Welt, *J. Gen. Physiol.* **63**, 351 (1974).
49. J. D. Carieres & J. C. Ellory, *J. Physiol.* **271**, 289 (1977).
50. J. G. Nørby & J. Jensen, *Biochim. Biophys. Acta* **233**, 104 (1971).
51. C. Hegevary & R. L. Post, *J. Biol. Chem.* **246**, 5234 (1971).
52. J. C. Skou, *Biochim. Biophys. Acta* **339**, 246 (1974).
53. R. L. Post, S. Kume & F. N. Rogers, Alternating paths of phosphorylation of the sodium and potassium pump of plasma membranes, in *Mechanisms in Bioenergetics*, G. F. Azzone, L. Ernster, S. Papa, E. Quagliariello & N. Siliprandi, Eds., Academic Press, New York, 1973, pp. 203–218.
54. S. J. D. Karlish, D. W. Yates & I. M. Glynn, *Nature* (Lond.) **263**, 251 (1976);
55. L. A. Beauge & I. M. Glynn, *Nature* (Lond.) **280**, 510 (1979).
56. P. J. Garrahan & A. F. Rega, *J. Physiol.* **223**, 595 (1972).
57. P. J. Garrahan, M. I. Pouchan & A. F. Rega, *J. Physiol.* **202**, 305 (1969).
58. W. F. Duddy & C. G. Winter, *Biochim. Biophys. Acta* **241**, 650 (1971).
59. R. C. Post, S. Kume, P. Tobin, B. Orcutt & A. K. Sen, *J. Gen. Physiol.* **54**, 306s (1969).

References

60. R. D. Keynes & R. A. Steinhardt, *J. Physiol.* **198**, 581 (1968).
61. E. Rogus & K. L. Zierler, *J. Physiol.* **233**, 227 (1973).
62. R. L. Birks & D. F. Davey, *J. Physiol.* **202**, 171 (1969).
63. E. E. Bittar, S. Chen, G. B. Danielson, H. Hartmann & E. Y. Tong, *J. Physiol.* **220**, 1 (1972).
64. E. J. Conway, R. P. Kernan & J. A. Zadunaisky, *J. Physiol.* **155**, 263 (1961).
65. R. P. Kernan, *J. Physiol.* **162**, 129 (1962).
66. R. Creese, *J. Physiol.* **197**, 255 (1968).
67. J. F. Manery, D. R. H. Gourley & K. C. Fisher, *Can. J. Biochem. Physiol.* **34**, 893 (1956).
68. L. B. Smillie & J. F. Manery, *Am. J. Physiol.* **205**, 67 (1960).
69. K. L. Zierler, *Am. J. Physiol.* **195**, 515 (1959).
70. R. Lorente de No, *Studies Rockefeller Inst. Med. Res.* **131**, 1 (1947).
71. C. M. Connelly, *Rev. Mod. Phys.* **31**, 475 (1959).
72. A. L. Hodgkin, *Proc. Roy. Soc. B* **148**, 1 (1958).
73. F. H. Shaw & S. E. Simon, *Nature* (Lond.) **176**, 1031 (1955).
74. J. A. Johnson, *Am. J. Physiol.* **187**, 328 (1956).
75. W. K. Stephenson, *J. Cell. Comp. Physiol.* **50**, 105 (1957).
76. R. P. Kernan, *Nature* (Lond.) **193**, 986 (1962).
77. R. P. Kernan, *Proc. Roy. Irish Acad.* **64**, 401 (1966).
78. R. P. Kernan & A. Tangney, *J. Physiol.* **172**, 32P (1964).
79. R. H. Adrian & C. L. Slayman, *J. Physiol.* **184**, 970 (1966).
80. E. J. Harris & S. Ochs, *J. Gen. Physiol.* **187**, 5 (1966).
81. L. J. Mullins & K. Noda, *J. Gen. Physiol.* **47**, 117 (1963).
82. S. B. Cross, R. D. Keynes & R. Rybova, *J. Physiol.* **181**, 865 (1965).
83. G. A. Kerkut & R. C. Thomas, *Comp. Biochem. Biophys.* **14**, 167 (1965).
84. R. C. Thomas, *J. Physiol.* **201**, 495 (1969).
85. R. P. Kernan, Electrogenic or Linked Transport, in *Membranes and Ion Transport*, Vol. 1, E. E. Bittar, Ed., Wiley, New York, 1970, pp. 395–431.
86. K. Koketsu, *Adv. Biophys.* **2**, 77 (1971).
87. J. M. Ritchie, *Current Topics Bioeng.* **4**, 327 (1971).
88. R. C. Thomas, *Physiol. Rev.* **52**, 563 (1972).
89. R. A. Sjödin & O. Ortiz, *J. Gen. Physiol.* **66**, 269 (1975).
90. M. Dockry, R. P. Kernan & A. Tangney, *J. Physiol.* **186**, 187 (1966).
91. R. P. Kernan, *Nature* (Lond.) **210**, 537 (1966).
92. R. P. Kernan, *J. Physiol.* **236**, 12P (1973).
93. E. X. Albuquerque, F. T. Schuh & F. C. Kauffman, *Pflügers Arch.* **328**, 36 (1971).
94. R. P. Kernan & I. McCarthy, *J. Physiol.* **226**, 62P (1972).

95. S. Locke & H. C. Solomon, *J. Exp. Zool.* **166,** 377 (1967).
96. H. Lorkovic & R. J. Tomanek, *Am. J. Physiol.* **232,** C109 (1977).
97. K. L. Zierler, *Am. J. Physiol.* **197,** 515 (1959).
98. K. L. Zierler, E. Rogus & C. F. Hazlewood, *J. Gen. Physiol.* **49,** 433 (1966).
99. P. C. Caldwell, *Physiol. Rev.* **48,** 1 (1968).
100. D. Erliz & S. Grinstein, *J. Physiol.* **259,** 13 (1976).
101. R. D. Moore, *J. Physiol.* **232,** 23 (1973).
102. W. A. Gavryck, R. D. Moore & R. C. Thompson, *J. Physiol.* **252,** 43 (1975).
103. E. T. Hays, T. M. Dwyer, P. Horowicz & J. G. Swift, *Am. J. Physiol.* **227,** 1340 (1974).
104. R. H. Evans & J. W. Smith, *J. Physiol.* **232,** 81P (1973).
105. E. M. Rogus, L. C. Cheng & K. Zierler, *Biochim. Biophys. Acta* **464,** 347 (1977).
106. N. Tashiro, *Br. J. Pharmacol.* **48,** 122 (1973).
107. A. P. Somlyo & A. V. Somlyo, *Fed. Proc.* **28,** 1634 (1969).
108. B. W. Festoff, K. L. Oliver & N. B. Reddy, *J. Membr. Biol.* **32,** 331 (1977).
109. T. Clausen & J. A. Flatman, *J. Physiol.* **270,** 383 (1977).
110. E. W. Sunderland, G. S. Robinson & R. W. Bulcher, *Circulation* **37,** 279 (1968).
111. A. H. Caswell, S. B. Baker, H. Boyd, L. T. Potter & M. Garcia, *J. Biol. Chem.* **253,** 3049 (1978).
112. M. Tada, M. A. Kirchberger, D. I. Repke & A. M. Katz, *J. Biol. Chem.* **249,** 6174 (1974).
113. M. L. Entman, G. S. Levey & S. E. Epstein, *Circ. Res.* **25,** 429 (1969).
114. M. L. Entman, M. A. Goldstein & A. Schwartz, *Life Sci.* **19,** 1623 (1975).
115. L. C. Chang, E. M. Rogus & K. Zierler, *Biochim. Biophys. Acta* **464,** 338 (1977).
116. Y. C. Hegevary, *Fed. Proc.* **35,** 835 (1976).
117. J. B. Li & L. S. Jefferson, *Am. J. Physiol.* **232,** E243 (1977).
118. E. Page & S. R. Storm, *J. Gen. Physiol.* **48,** 957 (1965).
119. K. R. H. Repke & H. J. Portius, *Experientia* **19,** 452 (1963).
120. G. A. Langer, *Fed. Proc.* **36,** 2231 (1977).
121. K. Temma, T. Akera, D. D. Ku & T. M. Brody, *Pharmacologist* **18,** 144 (1976).
122. T. Akera, M. K. Olgaard, K. Temma & T. M. Brody, *J. Pharmacol. Exp. Therap.* **203,** 675 (1977).
123. T. Akera, *Nature* (Lond.) **198,** 569 (1977).
124. R. Weingart, R. S. Kass & R. W. Tsien, *Nature* (Lond.) **273,** 389 (1978).
125. D. S. Miura, B. F. Hoffman & M. R. Rosen, *Fed. Proc.* **35,** 320 (1976).
126. J. H. Widdicombe, *J. Physiol.* **266,** 235 (1977).

127. A. F. Brading & J. H. Widdicombe, *J. Physiol.* **238**, 235 (1974).
128. R. A. Brace, *Proc. Soc. Exp. Biol. Med.* **145**, 1389 (1974).
129. K. Brecht, P. Konold & G. Gebert, *Physiol. Bohemoslov.* **18**, 15 (1969).
130. D. K. Anderson, *Fed. Proc.* **35**, 1294 (1976).
131. F. J. Brinley, Jr., & L. J. Mullins, *J. Gen. Physiol.* **50**, 2303 (1967).
132. F. J. Brinley, Jr., & L. J. Mullins, *J. Gen. Physiol.* **52**, 181 (1968).
133. F. J. Brinley, Jr., & L. J. Mullins, *Ann. N.Y. Acad. Sci.* **242**, 406 (1974).
134. P. DeWeer, *J. Gen. Physiol.* **56**, 583 (1970).
135. L. J. Mullins, Active Transport of Na^+ and K^+ Across the Squid Axon Membrane, in *Role of Membranes in Secretory Processes*, North-Holland, Amsterdam, 1972, pp. 181–202.
136. L. Lambotte, *J. Physiol.* **269**, 53 (1977).
137. B. Claret, M. Claret & J. L. Mazet, *J. Physiol.* **230**, 87 (1973).
138. P. M. Beigelman & L. J. Thomas Jr., *J. Membr. Biol.* **8**, 181 (1972).
139. E. J. Conway & W. McD. Armstrong, *Biochem. J.* **81**, 631 (1961).
140. E. J. Conway, P. F. Duggan & R. P. Kernan, *Proc. Roy. Irish Acad.* **63**, 93 (1963).
141. W. McD. Armstrong & A. Rothstein, *J. Gen. Physiol.* **48**, 61 (1964).
142. E. J. Conway & P. F. Duggan, *Biochem. J.* **69**, 265 (1958).
143. E. J. Conway & P. T. Moore, *Biochem. J.* **57**, 523 (1954).
144. E. J. Conway, *Int. Rev. Cytol.* **4**, 377 (1955).
145. E. J. Conway & R. P. Kernan, *Biochem. J.* **61**, 32 (1955).
146. P. A. Kometiani, Concerning the Role of the Redox Potential of the Outer Medium on the Distribution of Cellular Electrolytes, in *Membrane Transport and Metabolism*, A. Kleinzeller & A. Kotyk, Eds. Czechoslovak Academy Sci. Publ., Prague, 1961, pp. 180–192.
147. C. Reilly & F. Sherman, *Biochim. Biophys. Acta* **95**, 65 (1965).
148. C. Reilly, *Biochem. J.* **91**, 447 (1964).
149. L. Kovac, G. S. P. Groot & E. Racker, *Biochim. Biophys. Acta* **256**, 55 (1972).
150. M. Cockburn, P. Earnshaw & A. A. Eddy. *Biochem. J.* **146**, 705 (1975).
151. P. L. Werden, W. Epstein & S. G. Schultz. *J. Gen. Physiol.* **50**, 1641 (1967).
152. A. Seaston, G. Carr & A. A. Eddy, *Biochem J.* **154**, 669 (1976).
153. D. B. Rhoads & W. Epstein. *J. Gen. Physiol.* **72**, 283 (1978).
154. W. Epstein & S. G. Schultz. *J. Gen. Physiol.* **49**, 469 (1966).
155. F. M. Harold, *Curr. Topics Bioeng.* **6**, 83 (1977).

6
Potassium Fluxes in Mitochondria

Mitochondria, because of their special function in energy conservation in the cell and the possible role of electrolyte metabolism in this process, merit separate consideration in relation to potassium fluxes and distribution. Two questions that need consideration are whether these cations are passively distributed across the mitochondrial membranes[1] and, if not, whether their net movement may have anything to contribute to the primary function of these subcellular particles. The inner membrane of the mitochondria is said to have a very active proton pump which continually removes these cations from the inner matrix and in the process either takes up potassium ions in neutral chemical coupling or generates an inner negativity with which the potassium ions come into equilibrium by passive movement across the membrane. Potassium ions sequestered in the matrix compartment of isolated mitochondria seem to exchange only very slowly under normal conditions.[2] When the particles are immersed in a solution of 120 mM KCl, they show little change unless the potassium ionophore valinomycin is added to the suspension whereupon swelling occurs, the salt and water being taken up without the need for energy expenditure.[3]

Although mitochondria are usually surrounded by the cytosol, which is rich in potassium, they have on isolation been extensively studied while suspended in solutions containing little of this cation, which raises doubts as to the relevance of such studies to the operation of the particles *in situ*.

The dominant cation within freshly isolated mitochondria was found to be potassium, with over 100 m.mole·kg^{-1} particles, followed by magnesium with about 30 m.mole·kg^{-1} and by calcium with about 10 m.mole·kg^{-1} particles. Their sodium content was negligible. While their magnesium content remained remarkably constant, their retention of potassium and calcium was evidently dependent on metabolic energy, since when metabolism was blocked there was a slow loss of these cations. Accumulation of potassium occurred in potassium-depleted mitochondria even in the absence of ionophores, provided they were respiring in the presence of P_i or in medium at high pH. Untreated mitochondria readily took up potassium against steep concentration gradients when their membranes were made permeable to K^+ by the addition of valinomycin or similar neutral ionophores; and associated with this uptake was the appearance of hydrogen ions in the medium, enhanced respiration, and swelling of the particles.

Mitochondria also used ATP as energy source for the accumulation of potassium, 7 equivalents of the cation being taken up for each equivalent of ATP broken down. When oxidizable substrate was used as energy source, the uptake rates were increased, but then only about 3 equivalents of potassium were transported for each mole of ATP used. The ATP is probably not the immediate energy source for the transport process, since in the respiring mitochondria potassium uptake was not blocked by oligomycin which is an ATPase inhibitor. When valinomycin was added to mitochondria suspended in K-free media, the loss of potassium from the particles was accompanied by a generation of ATP.[4,5] So there was evidently some ATPase activity associated with active transport of potassium.

The main barrier to diffusion in mitochondria seems[6] to be the inner membrane (see Plate 9) since the outer membrane allows through solutes of molecular weight of around 1000. The inner membrane therefore appears to be responsible for an energy-linked transfer of potassium ions although the significance of such transfer is doubtful *in situ* where potassium concentration may be virtually the same inside and outside the particles.

THE MEMBRANE POTENTIAL IN MITOCHONDRIA

A problem in the measurement of the electrochemical potential of potassium and other ions in mitochondria has been the uncertainty about the

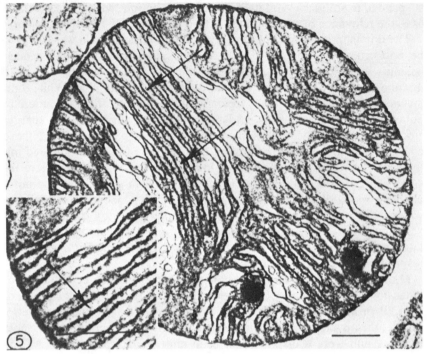

Plate 9 Mitochondrion fixed in 1% OsO_4. The intracristal structure is not observed after this fixation procedure. Note the electron-opaque material (arrows) in the intracristal space. Inset is a higher magnification (×12,500) of the electron-opaque material (×66,250). From J. D. Hall & F. L. Crone, *J. Cell Biol.* **48**, 420 (1971).

magnitude and polarity of the electrical charge across the inner membrane. The first direct measurements[7] were made in isolated mitochondria (3–4 μm in diameter) from the salivary glands of *Drosophila*. Potential measurements by microelectrodes were made in different metabolic states and in the presence of relatively low external potassium concentrations. In state 4 (slow respiration, high [ATP]/([ADP] + [P]$_i$) ratio), the mean membrane potential using a medium with 10 mM K was 9.8 ± 2.5 mV inside positive, changing in state 3 (fast respiration, medium [ATP]/([ADP] + [P]$_i$) ratio) to +19 mV approximately. It was also found that the potential in state 4 was unaffected by changes in the potassium concentration of the medium. For example, in the presence of 50 mM K, the membrane potential was +9.2 to +10.4 mV, which was virtually identical with that found in 10 mM K. Because the membrane potential also appeared to be insensitive to the addition of cyanide and 2,4-dinitrophenol even in state 3, it was concluded that the membrane potential played no significant role in the energy conservation by these particles.

Support for the view that the internal charge was positive also came from the measurement[8] of the distribution of ^{14}C-labeled anions and the application of the Nernst equation to the distribution. Similar experiments[7] in which anion distribution and associated diffusion potentials have been compared with measured potentials seemed to confirm that anions were passively distributed across the mitochondrial membrane of *Drosophila* and that the internal potential was about 10 mV positive. When osmotic swelling was induced by suspending the particles in hypotonic media (109 mOsm instead of 450 mOsm), the membrane potential fell by about 6.7 mV, which was close to the value predicted for an anion diffusion potential. Although E_K was not calculated under these experimental conditions, it would seem reasonable to suppose that its value would be at least -45 mV, suggesting that the potassium ions were far from equilibrium distribution across the mitochondrial membrane. It has been suggested that the microelectrodes may not have penetrated the inner membrane[9] and that such small particles could not support a true resting potential for more than a few msec due to leakiness induced by penetration of the electrode.[10,11]

In another series of experiments[12] the addition of 1.4×10^{-7} M valinomycin to the mitochondria in media containing 5 mM K caused the membrane potential to change from about $+17$ mV to about -29 mV, presumably owing to a diffusion of potassium ions down an electrochemical gradient and out of the particles. The induced potential here was maintained for at least 25 sec, which again raised doubts as to its true nature. One would expect here the type of decay of potential seen in erythrocytes penetrated by microelectrodes.[13] It was argued that the potential change measured in mitochondria following addition of valinomycin, which would be consistent with these particles normally having very low potassium permeability, might have been due to an artifact produced in the microelectrode. The tip of the electrode might have become blocked or coated with hydrophobic material and the addition of the valinomycin might have converted it to a potassium-selective electrode, the exit of K^+ from the inside of the electrode producing the observed potential change. It was reported however that microelectrodes filled with NaCl instead of KCl also recorded these electrical changes. Furthermore addition of valinomycin did not cause the internal potential to become negative when the mitochondria were suspended in media containing 60–80 mM K, which approaches the expected intracellular concentration of potassium under these conditions.

The conclusions from these experiments[7] were that the mitochondrial membrane potential was the result of the distribution of anions imposed by a Donnan effect, the potassium ions being far from equilibrium distribution, which would be in keeping with the evidence for low membrane permeability to the potassium ion and the energy-linked transport of this cation across the membrane.

The contrary view of the magnitude and polarity of the mitochondrial potential was based mainly on the measurement of the distribution of ^{42}K and ^{86}Rb between these particles and the suspending fluid[14-16] in the presence of the ionophore valinomycin, which was believed to facilitate the passive movement of these cations across the membrane and into equilibrium with the existing potential. The existing potential could then be determined by the use of the Nernst equation. The equilibrium potential for these cations was about -150 mV. The distribution of guanidinium ions, which apparently are not actively transported, was found[17] to be in equilibrium with a membrane potential of -60 to -30 mV. The discrepancy found between the positive and negative membrane potentials probably arose from the smallness of the particles under examination, from the variety of methods used for measurement, and from the deficiencies and dubious assumptions made in relation to their application in some cases.

In relation to the methods above that yielded negative values for membrane potential, the following question might be asked. Would the distribution of sodium ions measured across a muscle fiber membrane in the presence of a sodium ionophore be acceptable as a sound basis for calculating the normal resting potential of the fiber? The negative response to this question underlines the weakness of some of the methods that have been used and that are apparently also based on the assumption of a passive response within the particles. Confirming this point was the finding that the distribution of potassium in the presence of valinomycin, of sodium in the presence of gramicidin, and of tetrapropylammonium in the presence of tetraphenylboron all yielded different values for membrane potential[18] under steady-state conditions.

POTASSIUM IONS AND THE CHEMIOSMOTIC HYPOTHESIS

The interest in the magnitude of the mitochondrial membrane potential has arisen to a large extent because of the chemiosmotic hypothesis of Mitchell,[19,20] according to which energy needed for the generation of ATP comes from the electrochemical potential of protons across the inner mitochondrial membrane. This electrochemical gradient in turn is believed to result from the active transport of protons out of the mitochondrial matrix by a pump operating through the respiratory chain in a manner similar to that suggested by Conway much earlier to account for gastric acid secretion and the excretion of hydrogen ions by yeast cells fermenting in glucose in the presence of potassium chloride. The total protonmotive force ΔP is made up of electrical and chemical components and expressed as follows:

$$\Delta P = E - Z\, \Delta \text{pH} \tag{36}$$

where E is the mitochondrial membrane potential, ΔpH is the pH gradient across the membrane, and Z is approximately 59 at 30°C, with ΔP and E being expressed in millivolts.

When mitochondria were brought to control conditions (state 5) by equilibration in an anaerobic suspending medium, the protomotive force could be regarded as zero, but E° had then a value determined presumably by a Donnan or Nernst potential. Under these conditions,

$$\Delta P^0 = 0 = E^0 - Z\, \Delta \text{pH}^0 \tag{37}$$

In the presence of 100 μg valinomycin·g⁻¹ mitochondrial protein, E^0 was then calculated[14] from the distribution of potassium between the particles and suspending fluid using the Nernst equation. The effect of this antibiotic on the permeability of the mitochondrial membrane was then examined from the kinetics of potassium and proton fluxes following the application of a driving force or disequilibrium in the form of alkalinization of the external medium by the addition of choline hydroxide. On addition of valinomycin at concentrations of zero, 1, 10, and 100 μg·g⁻¹ protein, total membrane conductance and potassium conductances (in brackets) were 2.9 (0), 9.1 (6.2), 24.6 (21.7), and 40,0 (37.1) μg·equiv·charge·sec⁻¹mV⁻¹·g⁻¹ protein, respectively. The conductance changes induced by the ionophore were not proportional to the quantity added, so perhaps higher conductance rates for potassium were limited by the rate of exit of protons from the mitochondria in the exchange process. Since the redistribution of potassium ions across the mitochondrial membrane following external alkalinization seemed to be complete within 5.5 sec in media containing 10 or 1 mM K at the highest concentration of ionophore used and only 20 sec under less favorable conditions, it was suggested that an activation of active transport of potassium did not take place.

Having measured permeability under conditions where metabolic changes were unlikely to occur, the rates and the extent of change in potassium distribution were then examined during respiration by means of external electrodes selective for potassium ions and for protons. Pulses of respiration lasting 200 sec were induced in mitochondria suspended in media containing 150 mM choline chloride, 3 mM glyclyglycine, 0.2 mM EGTA (to prevent backlash associated with looseness of coupling[21]), and 100 μg valinomycin·g^{-1} protein, with 2 mM β-hydroxybutyrate as substrate. In some cases oligomycin (1 mg·g^{-1} protein) was added to block ATPase. It was found (Fig. 6.1) that under these conditions an outward translocation of protons coupled to an equivalent entry of potassium ions took place within these particles. In the absence of oligomycin however there was a deficiency of protons in the medium as the mitochondria went into state 4, and this was greatly enhanced in the presence of endogenous ADP and P_i. It was suggested that the deficiency was due to protons being reabsorbed by the mitochondria in the synthesis of ATP, a process blocked by oligomycin. In state 4, while the protonmotive force seemed to remain constant at about 227 mV, under different conditions E seemed to change from a high of -199 mV in K-depleted mitochondria in low-K medium (pK or $-\log [K]_0$ = 4.6–4.8) to a low of -83 mV in normal mitochondria immersed in a similar medium but with 10 mM K.

The contribution of membrane potential to the electrochemical gradient for hydrogen ions ranged therefore from 87 to 36%, depending on potassium concentrations inside and outside the mitochondria. The significance of the potassium movements across the mitochondrial membrane is that such an electrophoretic flux could lead to the interconversion of E and ΔpH.[22] The maximum $[K]_i/[K]_0$ ratio maintained by respiring mitochondria from liver was about 1600, corresponding to an equilibrium potential of -192 mV.[23] When this ratio exceeded 2000, the addition of valinomycin caused an efflux of potassium rather than influx. Because the presence of valinomycin did not seem to stimulate respiration rate in state 4, it was argued that the unequal distribution of potassium ions across the membranes of the cristae could not be due to a metabolically driven potassium pump but that these ions, in spite of their obvious low permeability, were maintained in equilibrium with potential. However others[24] believed that the ionophore activated a potassium pump driven either by respiration or by hydrolysis of ATP and that this transport was responsible for the unequal distribution of this cation. It was noted however[14] that the initial coupling ratio of proton efflux over potassium influx increased when the concentration of valinomycin in the medium was low; this was taken to indicate that the latter process was dependent on membrane permeability.

In calculating E from the steady-state distribution of potassium between mitochondria and the medium, there was uncertainty about the quantity of solvent water in the matrix and the possible inequality of activity coefficient for potassium inside and outside the particles.

The amount of potassium in state 4 mitochondria has been estimated[23] at about 80 nM·mg^{-1} protein and the matrix volume at 0.4 μl·mg^{-1} protein

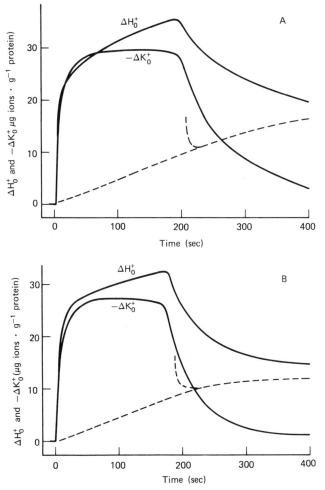

Figure 6.1 Time course of ΔH_0^+ and ΔK_0^+ during respiratory pulses in mitochondrial suspensions at 25°C under the conditions mentioned in the text. A, in absence of oligomycin; B, 1 mg of oligomycin per g mitochondrial protein added. Broken curves indicate ΔH_0^+ baselines obtained by adding 1 μM FCCP immediately after the end of the respiratory pulse. From P. Mitchell & J. Moyle *Eur. J. Biochem.* **7,** 471 (1969).

in 250 mOsm medium. This was equivalent to $[K]_i$ of 200 m.mole·kg^{-1} mitochondrial water. An independent estimate of $[K]_i$ was reached on the assumption that osmotic equilibrium was present and that most of the osmotic activity provided by the matrix cation depended on potassium. Its osmotic activity would then be equal to at least half the osmolarity of the medium, or 120 mM in 250 mOsm medium.

In earlier studies[24] $[K]_i$ of mitochondria in the presence of valinomycin was estimated by three procedures to be 60–80 mM, which was much lower than the osmotic equivalent of the reaction medium (170 mM). In the absence of direct measurement of membrane potential here, the evidence for an internal negativity of the mitochondrial matrix based on equilibrium with the distribution of potassium ions is necessarily indirect and circumstantial. While the constancy of the apparent protonmotive force ΔP and consequently the reciprocal relationships between $-z\,\Delta\mathrm{pH}$ based on direct measurement on lysed mitochondria and E calculated from potassium distribution were certainly suggestive of a passive distribution of the latter between the matrix and external fluid, contrary evidence still remains considerable.

For example, membrane potential has also been determined from the unidirectional fluxes of potassium ions across the mitochondrial membrane.[25] In accordance with the predictions of the Goldman constant field equation, the net flux of potassium ions across a permeable membrane may be defined as follows:

$$m_K^i - m_K^o = \frac{P_K FE}{RT}\left[\frac{[K]_o - [K]_i\, e^{FE/RT}}{1 - e^{FE/RT}}\right] \quad (38)$$

from which unidirectional fluxes may be expressed as

$$m_K^i = \frac{P_K\,[K]_o(FE/RT)}{1 - e^{FE/RT}} \quad (39)$$

and

$$m_K^o = \frac{P_K[K]_i\,(FE/RT)\,e^{FE/RT}}{1 - e^{FE/RT}} \quad (40)$$

The flux rate constants $k_i = m_K^i/[K]_o$ and $k_o = m_K^o/[K]_i$ were then calculated from the experimental data and the theoretical membrane potentials calculated using the equation

$$E = -\frac{RT}{F}\ln\frac{m_K^i[K]_i}{m_K^o[K]_o} = -\frac{RT}{F}\ln\frac{k_i}{k_o} \quad (41)$$

that is, from the ratio of the rate constants. This yielded potentials of -70 to -80 mV, which were considerably less than those reported elsewhere.[14,23] As these measurements were made in the absence of ionophores and much higher values were found in the presence of ionophores, this might suggest that potassium ions were far from equilibrium with membrane potential in the former conditions.

If, as has been claimed, potassium ions move into the mitochondria passively under the influence of a potential generated by an electrogenic proton pump, then a comparison of unidirectional potassium fluxes in the presence and absence of metabolism might be expected to reflect the electrophoretic potassium movement. This did not seem to be the case, for although the rate constant for potassium influx, k_i, decreased when metabolism was blocked, that of efflux k_o did not increase as predicted but in fact m_K^o actually decreased significantly in the presence of sodium cyanide and antimycin A.

Initially this observation was made under conditions where ^{42}K uptake decreased without apparent change in intramitochondrial potassium level, so that it had to be assumed that K efflux had also decreased. However the dependence of potassium efflux on respiration was subsequently confirmed[26] in separate experiments in which ^{42}K efflux was measured directly in the near-steady state and found to be increased significantly by respiration and to be sensitive to uncouplers and exogenous magnesium. The apparent confirmation of an energy-dependent efflux as well as an energy-dependent influx was not consistent with electrophoretic potassium uptake by mitochondria but was understandable in the context of the normal environment of mitochondria *in situ* where the external potassium concentration may be over 140 mM. It was also found that a number of factors which influenced ^{42}K influx under steady-state conditions had a parallel effect on ^{42}K efflux. Both ^{42}K fluxes, for example,[27] as well as requiring respiration were activated by P_i and by mersalyl (10 n.mole·mg^{-1} protein), which is a blocker of P_i and dicarboxylate transport. This presented a rather anomalous situation which apparently has not been resolved. The influx of ^{42}K in the presence of P_i was strongly inhibited by the addition of ADP which initiated phosphorylation. This was the opposite response to that obtained[24] with mitochondria in state 4, where a loss has been observed in the presence of 1 mM K_o. Addition of oligomycin here reversed the direction of K movement by blocking the interaction of the ATPase with the K gradient. The inward K flux was then coupled to respiration, since subsequent addition of rotenone to block this process again led to net loss of potassium, which was increased further if valinomycin was added to increase K permeability.

The respiration-dependent efflux of ^{42}K from heart mitochondria was not activated by P_i in a NaCl medium as opposed to a KCl medium, which would suggest that it involved K-K exchange on a carrier. However conditions which promoted net entrance of sodium into the particles, namely, addition of EDTA which removed Mg from mitochondrial membranes activated such efflux. Exogenous ATP could not replace respiration as energy source for ^{42}K efflux. The maximum phosphate-dependent ^{42}K efflux from heart mitochondria occurred in the presence of 2–5 mM P_i and 100 mM KCl. This efflux was inhibited by 5 mM Mg only if this cation had been omitted from washing solution following labeling of the particles, while ^{42}K influx was competitively inhibited by concentrations up to 5 mM Mg at least without such omission. The addition of P$_i$ to ^{42}K-labeled heart mitochondria respiring in 100 mM KCl resulted in the accumulation of 60–70 nM K$^+$·mg^{-1} protein with swelling but with no loss of ^{42}K initially. The replacement of potassium chloride by potassium acetate in the medium led to an even greater swelling and potassium uptake. The swelling of mitochondria which probably results from a Donnan effect is most puzzling in the case of phosphate movements. Respiring particles in solutions containing 100 mM K phosphate became quite swollen and their overall ^{42}K efflux greatly suppressed, in contrast to those in 100 mM KCl and 2–5 mM P$_i$. In the latter, mersalyl eliminated K influx while inducing large efflux of ^{42}K. There was generally no increase in ^{42}K efflux and even some retention when net uptake of K was occurring in potassium acetate solution.

These results have been interpreted in terms of a chemiosmotic coupling model (Fig. 6.2), in which potassium ions were transported into the mitochondria through a low-affinity potassium uniport evident in the K–K exchange properties of the carrier, and with its exit largely regulated through a potassium–proton exchanger. Respiration was believed to be responsible for the establishment of a pH gradient across the mitochondrial membrane, which could then be converted into a potential gradient

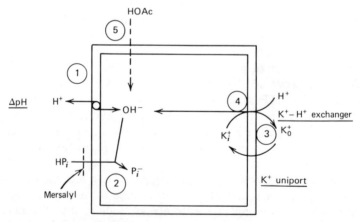

Figure 6.2 Proposed chemiosmotic model for turnover of potassium in heart mitochondria. 1, Production of ΔpH by respiration. 2, Anion transporter converting part of ΔpH to E. 3, A voltage-gated uniport of potassium permits influx when E exceeds a limiting value; this is the pathway for net K uptake (K$_m$ = 12 mM). 4, K$^+$/H$^+$ exchanger activated when ΔpH exceeds a limiting value, also the pathway for net extrusion. 5, In the presence of acetate the equilibrium concentration of free acetic acid converts ΔpH into an acetate gradient E_a. From E. Chavez, D. W. Jung & G. P. Brierley, *Arch. Biochem. Biophys.* **183**, 460 (1977).

E through the operation of a phosphate transporter and other anion exchangers present in the membrane. On the one hand the OH^-–anion exchanger created the conditions favoring potassium uptake, while K^+–H^+ exchange under conditions where ΔpH had become sufficiently great promoted K efflux. The mersalyl-induced ^{42}K efflux could then be explained by its ability to inhibit the conversion of ΔpH to E. The inhibition of the P–OH^- exchanger in this case by the sulfhydryl reagent N-ethylmaleimide (NEM) has been used[28] in the promotion of K^+ influx through possible K^+ and OH^- cotransport in rat liver mitochondria.

In these experiments unidirectional K^+ influx showed saturation kinetics and could be driven by a pH_0-dependent mechanism. Increase of pH_0 from 7.0 to 8.0 raised the apparent V of K^+ influx without significant alteration in K_m for K^+. The NEM and also high external $[P]_i$ stimulated K^+ influx at alkaline pH by preventing P_i efflux in exchange for OH^- influx, the latter being apparently coupled then with K^+ influx.

In summary, then, the finding that the steady-state influx of ^{42}K into rat liver mitochondria required respiration, was sensitive to uncouplers, showed saturation kinetics, and was competitively inhibited by Mg^{2+} seemed strongly to support the view that uptake of this cation was carrier mediated and energy linked. The observation that both influx and efflux of the cation were increased or decreased in a parallel manner by a number of factors seemed to indicate a potassium uniport. The existence of a K^+–H^+ exchanger operating in parallel with the K pump has also been postulated because of the sensitivity of the potassium fluxes to external pH. Through this exchange K^+ could be extruded from the matrix in exchange for the H^+ taken up.

Whatever the model used, it must account for the variability of the H^+–K^+ exchange ratio and the tendency for mitochondria to swell in the presence of external potassium and permeable anions. The latter would seem to be a Donnan effect. Use has been made of the latter to demonstrate an outwardly directed potassium (and sodium) transport which apparently controls the volume of the mitochondria.

VOLUME CONTROL AND POTASSIUM TRANSPORT IN MITOCHONDRIA

The presence of succinate in the medium caused the H^+–K^+ coupling ratio to decrease, apparently because of entry of potassium accompanied by the anion rather than in exchange for protons. The ratio here fell from 1 to 0.2, which could be quantitatively accounted for by the accumulation of succinate. This suggested that active K^+ uptake[29] could cause the coupled movement of substrate anion and that it was not a case of proton extrusion drawing potassium passively into the matrix for electrical equilibrium. When uptake of succinate was inhibited at alkaline pH, the ratio increased. It seems likely that the swelling seen in mitochondria when succinate or other permeable anions were present in the medium was mediated by anion–OH^- exchange across the membrane. The swelling here paralleled the uptake of the anion.

When mitochondria were suspended in 120 mM KCl in a low-energy state and valinomycin was added, the particles accumulated the salt and water and swelled as a result. However when energy was then supplied, a net extrusion of KCl and water took place evidently through active transport.[3] Isolated beef heart mitochondria in 100 mM sodium or potassium nitrate at pH above 8 also swelled spontaneously at 35°C. They could be made to contract again in the energy-dependent reaction when respiration was initiated or ATP supplied.[30] The rate and extent of the contraction was greater in case of the sodium than the potassium salt and showed a strong pH dependence with an optimum pH of 6.8. Since mitochondrial K^+ was retained after massive sodium acetate accumulation, Na–K exchange was apparently not involved although matrix K may regulate the Na–H exchange.

The efficiency of contraction[31] (light absorbance change per O_2 or ATP consumed) decreased with increase of external pH and increased in the presence of Mg^{2+}, and this was believed to be due to modification of electrophoretic cation permeability controlling the simultaneous influx of salt and water. The osmotic contraction model invoked here postulated that a metabolism-dependent pH gradient could be used by the cation–H^+ exchanger to bring H^+ into the matrix, extruding sodium or potassium ions in the process. Since the H^+ would be neutralized in the alkaline interior of the particle, the exchange resulted in net loss of salt and water.

Another kind of shrinkage or contraction was seen in the mitochondria which had been allowed to swell slightly in the presence of potassium and phosphate ions. These apparently decreased in volume when ADP was added to the suspension, while there was the simultaneous inhibition of ^{42}K influx already mentioned until phosphorylation ceased. The net efflux of potassium associated with the shrinkage probably operated through the K^+–H^+ exchanger or may have been directly coupled to the synthesis of ATP from ADP and P_i. Oligomycin blocked this efflux and also reversed the inhibition of ^{42}K influx.

POTASSIUM FLUXES AND UNCOUPLING AGENTS

Compounds which uncouple oxidative phosphorylation through their facilitation of proton transport through membranes, including DNP[32] and

FCCP,[33] may also bring about an exchange of potassium and hydrogen ion in mitochondria. These uncoupling agents allow respiration to proceed at an optimal rate without synthesis of ATP and even in the presence of a high [ATP]/[ADP] ratio. All are weak acids with coupled π-electron systems. The protonated forms are lipid soluble and can therefore freely penetrate the membrane; but, in addition, in their deprotonated anionic form the delocalization of their charge by conjugation with the π-electron system makes possible their movement through the lipid phase.

It has been proposed[17,18] that energy conservation in mitochondria, which involves an interaction between the respiratory chain and the synthesis of ATP, may be mediated through the generation of an electrochemical gradient for hydrogen ions across the mitochondrial membrane. If this is true, then the action of uncoupling agents in increasing the permeability of membranes to H^+ may be understood if they allow the protons to leak down electrochemical gradients, thereby short-circuiting the proton pump. Indeed increased conductivity of artificial lipid bilayers by DNP which would be consistent with this view has been reported.[32] Such properties have also been found in other membranes on treatment with uncoupling agents including FCCP. The movement of protons across the mitochondrial membrane on treatment with uncoupling agents has been found to be accompanied by net fluxes of other ions, as, for example, a release of K^+ on addition of DNP.[34] Some have found K exit to occur only in the presence of low external pH. At low pH, DNP produced an extrusion of K^+ in exchange for protons which could be blocked by magnesium, suggesting carrier-mediated transport.[34] The uncoupler FCCP also produced an appreciable extrusion of K^+ from the particles but only if P_i was present in the medium, and in this case little uptake of protons seemed to take place.

When mitochondria were incubated in a K-free medium containing P_i with glutamate and malate as substrates, a substantial loss of potassium occurred when FCCP was added, while simultaneous changes in the redox state of cytochrome c and a rise in pH of the medium took place. When external phosphate was replaced by chloride however, although respiration and the respiratory chain were similarly affected, the exit of potassium from the particles was minimal, indicating a lack of correlation between K efflux and its influence on respiration.

When nigericin was added after FCCP in the presence of P_i, there was a small additional K loss, indicating that equilibrium had not been reached. But in the presence of Cl^- the antibiotic produced a dramatic K loss, making the total K exit comparable in both cases. Acetate could not substitute for P_i in facilitating K^+ exit in the presence of FCCP. Mg^{2+} inhibited the exit of K^+ when P_i was present but did not affect either respiration or pH. If P_i was added prior to FCCP, substantial potassium release occurred; but if added after, no movement of potassium was observed, even though P_i produced increased respiration and reduction of cytochrome c. This indicated the transitory nature of the K permeability change in the presence of FCCP. The exit of potassium in the presence of FCCP depended therefore on P_i–OH^- exchange or on K^+–H^+ exchange, the latter being facilitated in the presence of the antibiotic nigericin and the former in the presence of external P_i. The nigericin here differs from the other potassium ionophore valinomycin in one important respect, namely, that it also confers proton permeability on membranes, and this has been revealed by work on artificial lipid barriers.[35] It transfers hydrogen ions in a neutral complex and can then exchange these for potassium ions. The complex of potassium with nigericin is unchanged, while that with valinomycin is charged and can therefore confer conductance on the membrane. The latter can therefore more readily participate in energy-linked transport, being influenced, for example, by membrane potential.

In spite of an extensive release of potassium ions from mitochondria induced by nigericin, this monobasic ionophore did not inhibit oxidative phosphorylation in low concentration. Used in combination with valinomycin however, it did so synergically through a short-circuiting mechanism.[35] While the use of nigericin should bring about the movement of both protons and potassium ions down their electrochemical gradients with resulting shuttling of these ions across the pump-leak pathways in the membrane, the membrane potential may remain uncharged and potassium ions will equilibrate with this potential. Addition of a charge carrier in the form of valinomycin may discharge the membrane, leading to further potassium loss from the mitochondria. On the other hand, the ability of valinomycin to cause dissipation of the electrochemical gradients is limited by its inability to promote leakage of protons. The synergic actions of these two ionophores is therefore understandable and their action as uncouplers emphasizes the importance of the concentration gradients of potassium and hydrogen ions across the mitochondrial membrane in energy conservation within the particle.

The dual action of valinomycin in mitochondria should also be stressed at this point. In high concentration this ionophore seemed to short-circuit the energy-linked pumping of potassium ions across the membrane with dissipation of energy through increased shuttling of ions in and out of the particles. At low concentrations however, a transport of potassium ions into the mitochondria was promoted by valinomycin, coupled with increase in both respiration rate and in phosphorylation of endogenous or added nucleotide.[36] The P/O ratio here tended to be constant in spite of an increase in respiration that was even greater than that seen in the presence of the uncoupler DNP. If a "classical uncoupling agent" such as DNP or FCCP was added after the valinomycin,[35] it reversed the energy-linked potassium uptake produced by the latter, producing a much greater potassium release than seen in the presence of uncoupler alone.

Nigericin has been shown[37] to mimic the action of uncouplers here, blocking uptake of potassium induced in rat liver mitochondria by valinomycin while causing a rapid release of potassium previously accumulated. Valinomycin may therefore have a dual action, namely, to raise the permeability of the mitochondrial membrane and also to increase the affinity of an inwardly directed pump for potassium. This ionophore increased the turnover of potassium in the membrane, but when mitochondria were incubated in 0.23 mM K under which conditions the cation tended to be lost, the ionophore reversed the direction of its net flux.

The ionophore gramicidin, which seems to form ion-conducting channels in artificial lipid membranes through which potassium and other cations can pass, did not mimic the actions of valinomycin in mitochondria. It has been suggested that when isolated mitochondria were suspended in media having lower potassium concentrations than found *in situ*, some locus within the particle might have been deficient in this cation and that the deficit might be corrected by the energy-linked transport initiated by valinomycin. In the absence of potassium, valinomycin not only failed to stimulate oxidative phosphorylation but depressed it slightly, presumably due to energy dissipation in the cyclic movement of endogenous potassium. About 0.2 to 0.5 mM K could be detected in the medium by the K-electrode under these conditions.[38] The fact that valinomycin could promote phosphorylation while the quasi-ionophore gramicidin was without effect indicated that the mere facilitation of potassium movement through the mitochondrial membrane was insufficient of itself to stimulate the process.

Submitochondrial particles prepared from sonically disrupted mitochondria were used in studies of ionophore-induced cation movements across their membranes.[39] Although these particles can still perform oxidative phosphorylation and seemed to provide a simpler system in which to study the primary event in energy conservation, the sidedness of their membranes seem to be opposite to that of the mitochondria. In the presence of valinomycin and nigericin, these particles are also uncoupled, although neither antibiotic alone in the presence of potassium is effective. Nigericin stimulated K^+ influx while the subsequent addition of valinomycin caused the release of the potassium taken up. An uncoupling produced by NH_4^+ was associated with H^+ uptake and the collapse of the pH gradient set up by respiration. Addition of valinomycin subsequently mediated an efflux of the NH_4^+ taken up apparently electrophoretically.

REFERENCES

1. H. Rottenberg, *J. Membr. Biol.* **11**, 117 (1973).
2. E. J. Harris, J. D. Judah & K. Ahmed, *Adv. Bioeng.* **1**, 255 (1966).
3. A. Azzi & G. F. Azzone, *Biochim. Biophys. Acta* **131**, 468 (1967).
4. R. S. Cockrell, E. J. Harris & B. C. Pressman, *Nature* (Lond.) **215**, 1487 (1967).
5. E. Rossi & G. F. Azzone, *Eur. J. Biochem.* **12**, 319 (1970).
6. G. P. Brierley & D. E. Green, *Proc. Nat. Acad. Sci. U.S.* **53**, 73 (1965).
7. J. T. Tupper & H. Tedeschi, *Science* **166**, 1537 (1969).
8. E. J. Harris & B. C. Pressman, *Biochim. Biophys. Acta* **172**, 66 (1969).
9. H. Rottenberg, *J. Bioeng.* **7**, 61 (1975).
10. U. V. Lassen, A. M. Nielsen, L. Pape & O. O. Simonsen, *J. Membr. Biol.* **6**, 269 (1971).
11. U. V. Lassen, L. Pape, B. Vestergaard-Bogind & O. Bengtson, *J. Membr. Biol.* **18**, 125 (1974).
12. B. L. Maloff, S. P. Scordilis & H. Tedeschi, *Science* **195**, 898 (1977).
13. U. V. Lassen, *Proc. 1st Eur. Biophys. Congr.*, Vol. III, 1971, p. 13.
14. P. Mitchell & J. Moyle, *Eur. J. Biochem.* **7**, 471 (1969).
15. E. Padan & H. Rottenberg, *Eur. J. Biochem.* **40**, 431 (1973).
16. D. G. Nicholls, *Eur. J. Biochem.* **50**, 305 (1974).
17. T. Fields & B. C. Pressman, *Biophys. J.* **15**, 68a (1975).
18. S. Massari & T. Pozzan, *Arch. Biochem. Biophys.* **173**, 332 (1976).
19. P. Mitchell, *Nature* (Lond.) **191**, 144 (1961).
20. P. Mitchell, *Chemiosmotic Coupling and Energy Transduction*, Glynn Research, Bodmin, Cornwall, 1968.
21. P. Mitchell & J. Moyle, *Biochem. J.* **105**, 1147 (1967).
22. V. Skulachev, *FEBS Lett.* **74**, 1 (1977).
23. E. Rossi & G. F. Azzone, *Eur. J. Biochem.* **7**, 418 (1969).
24. E. J. Harris, G. Catlin & B. C. Pressman, *Biochem. J.* **6**, 1360 (1967).
25. J. J. Diwan & H. Tedeschi, *FEBS Lett.* **60**, 176 (1975).
26. E. Chavez, D. W. Jung & G. P. Brierley, *Arch. Biochem. Biophys.* **183**, 460 (1977).
27. D. W. Jung, E. Chavez & G. P. Brierley, *Arch. Biochem. Biophys.* **183**, 452 (1977).
28. J. J. Diwan & P. H. Lehner, *Membr. Biochem.* **1**, 43 (1978).
29. C. Rossi, A. Scarpa & G. F. Azzone, *Eur. J. Biochem.* **6**, 3902 (1967).
30. G. P. Brierley, M. Jurkovitz & E. Chavez, *Biochem. Biophys. Res. Commun.* **74**, 235 (1977).

31. G. P. Brierley & M. Jurkovitz, *Biochem. Biophys. Res. Commun.* **68,** 82 (1976).
32. J. Bielowski, T. E. Thompson & A. L. Lehninger, *Biochem. Biophys. Res. Commun.* **24,** 948 (1966).
33. P. Mitchell & J. Moyle, *Biochem. J.* **104,** 588 (1967).
34. J. D. Judah, A. E. M. McLean, K. Ahmed & G. S. Christie, *Biochim. Biophys. Acta* **94,** 441 (1965).
35. B. C. Pressman, E. J. Harris, W. S. Jagger & J. H. Johnson, *Proc. Natl. Acad. Sci. U.S.* **58,** 1949 (1967).
36. M. Höfer & B. C. Pressman, *Biochem.* **5,** 3919 (1966).
37. S. N. Graven, S. Estrado-O & H. A. Lardy, *Proc. Natl. Acad. Sci. U.S.* **56,** 654 (1966).
38. E. J. Harris, J. R. Cockrell & B. C. Pressman, *Biochem. J.* **99,** 200 (1966).
39. M. Montal, B. Chance & C. Lee, *J. Membr. Biol.* **2,** 201 (1970).

7
Potassium Ions in Cell Metabolism

A classic example of the activation of an enzyme system by cations is that found during muscle contraction when calcium ions entering the sarcoplasm reverse the inhibitory action of tropomyosin on the interaction of actin and myofilaments, with the consequent activation of the ATPase activity of the latter.[1] The mechanism of action here is fairly well understood, as are other cases of activation by divalent cations. The same cannot be said of activation of enzyme systems by monovalent cations, although it is now generally accepted that it occurs. One of the earliest reactions studied in this context was the phosphotransferase system operating between ATP and pyruvate. With the purification of the enzyme concerned with the reaction

$$\text{phosphopyruvate} + \text{ADP} \rightarrow \text{pyruvate} + \text{ATP} \qquad (42)$$

it seemed[2] that this reaction was irreversible, proceeding only in the direction indicated. It was suggested[3] that the apparent irreversibility of the reaction might be due to the absence of potassium from the enzyme system following the dialysis used in purification. This was confirmed subsequently[4] with the finding that pyruvate could be phosphorylated in the presence of phosphoglyceric acid, provided that potassium ions were also present in the medium. The enzyme concerned with the reaction, pyruvate kinase, bound one equivalent potassium for each equivalent pyrophosphate bound.[5] A similar stoichiometry was found in the case of fructokinase of ox liver, another enzyme requiring this cation. Since these early studies the number of cases of activation of monovalent cations found has increased considerably (Table 7.1).

The significance of intracellular potassium in the general metabolism of the cell was illustrated by the fact that yeast cells grown under conditions of potassium deficiency were smaller and irregular in shape in addition to having a respiration rate only 30% of normal. Growth rate was also markedly reduced in such cells, which had 98% of their potassium replaced by sodium ions.[6]

Table 7.1 Some Enzyme Systems Requiring Potassium Ions for Activation

Enzyme System	Source	Reference
Oxaloacetate decarboxylase	*Aspergillus niger*	48
ATP:D-fructose-1 phosphotransferase	Bovine liver	49
ATP:D-fructose 6-phosphate-1 phosphotransferase	Rabbit muscle	50
	Sheep brain	51
	Rat brain	52
Tryptophanase	*Escherichia coli*	53
L-Serine hydrolase	Rat liver	54
L-Threonine hydrolase	Sheep liver	54
S-Adenosylmethionine Alkyltransferase (cyclizing)	Rabbit liver	55
Aldehyde:NAD(P) oxidoreductase	Yeast	56
L-Malate:NAD(P) oxidoreductase	E. coli	57
ATP:Protein phosphotransferase	Brain	58
Carbamyl phosphate (acetylphosphate) phosphohydrolase	Guinea pig kidney	59
	Guinea pig brain	60
5,6,7,8-Tetrahydrofolate: NADP oxidoreductase	Guinea pig liver	61
L-Tyrosine:t-RNA ligase	Rat liver	62
	Pancreas	63
Propionyl-CoA:CO$_2$ ligase	Bovine liver	64
	Pig heart	65
ATP:Carbamyl phosphotransferase	Rat liver	66
Acetate:CoA ligase	Rat heart	67, 68
	Rabbit heart	69

POTASSIUM IN PROTEIN SYNTHESIS

It has been suggested that membrane-bound polyribosomes might be the major site of protein synthesis in bacteria *in situ*. Membrane-bound ribosomes have been isolated from cell lysates and associated with this "hot trichloracetic insoluble" material, labeled proteins have been detected following incubation of cells with labeled amino acids.[7] Furthermore *in vitro* studies have shown that membrane-bound ribosomes appeared to incorporate amino acids to a greater extent than free ribosomal material. It was found[8] subsequently during amino acid incorporation by cell-free preparations from gently lysed cells of *Bacillus amyloliquifaciens* in the presence of Tris buffer and magnesium ions at pH 7.6 that the distribution of ribosomal material between the soluble and membrane fractions was greatly influenced by the potassium concentration of the medium in the range of 0 to 100 mM K. In the absence of potassium about 37% of the ribosomal material was found to be associated with the membranes and was not removed by repeated washings with lysing buffer. However when the potassium concentration of the buffer was increased to 100 mM, the amount of ribosomal material solubilized increased until only about 5% remained attached to the membranes. In the absence of potassium, washed membranes containing no detectable ribosomal material reabsorbed some once more but only to the extent of about 50% of that found attached at a particular potassium concentration on extraction of freshly lysed cells. However when less than 10% of the ribosomes seemed to be attached to membranes, maximum incorporation of amino acids could still be demonstrated.

When preparations isolated from log phase cells of *B. subtilis*[9] containing polyribosomes in high proportion were examined in relation to optimal potassium concentration for protein synthesis in the presence of 0.01 M Mg^{2+} and 0.05 M Tris buffer at pH 7.6, this was found to lie in the range of 70 to 100 mM. At the lowest potassium concentration used, namely, 10 mM, the level of incorporation was about 50% of the maximum level and there was a decline to 66% of maximum when potassium was increased to 200 mM, at which concentration there was a loss of fast sedimenting particles and a significant and irreversible increase in the proportion of ribosomal subunits. The potassium therefore appeared to act in two ways, namely, through facilitating amino acid incorporation over one range of concentration while influencing the structural integrity of the ribosomes over the higher concentration range.[10]

The polymerization of ribosomes appeared to be influenced by the presence of potassium and other monovalent cations in the medium.[11-13] But while concentrations of potassium at or above 200 mM reversed polymerization, the complete absence of potassium seemed to have a similar effect. In K-free media, for example, monomeric endogenous inactive ribosomes from dormant invertebrate cells lost their ability to use poly-U as an artificial template and tended to dissociate irreversibly into subunits in spite of the stabilizing effect of Mg^{2+}. On the other hand endogenous active ribosomes were less sensitive to potassium lack, indicating the stabilizing influence of other components of the polyribosome complex.[11,12] This effect of potassium deprivation in converting ribosomal monomers into particles of lower sedimentation rate and higher sensitivity to ribonucleases was also found in material from rabbit reticulocytes and Ehrlich ascites tumor cells. These particles, while only slightly dissociated, lost their ability to incorporate phenylalanine on a poly-U template during potassium lack. Unlike the free monomers, polyribosomes seemed more resistant to the effects of K deficiency, retaining a great part of their ability to incorporate amino acids.

The potassium ions therefore appeared to act in two ways. First, at high concentration they may have screened negatively charged groups which participated in bridge formation in the linking of ribosomes to membranes through magnesium ions. At much lower concentrations the potassium ions appeared to act synergically with Mg^{2+} in stabilizing ribosomal structure. The concentration range for the latter effect was 0.5 to 1 mM, the cations NH_4^+, Rb^+, and Tl^+ also being effective substitutes for potassium in this respect. The finding that exposure to K-free media at 0°C of a few hours' duration was needed for structural alteration of ribosomes was attributed to the conformational importance of some potassium which might be firmly bound to specific sites in the ribosome structure. The effect of K deficiency was more rapid on ribosomal monomers than on polyribosomes in which the potassium might have been better screened.

Although it has been demonstrated that polymerization of ribosomes and amino acid incorporation are K sensitive in cell-free preparations and this in itself may explain potassium requirement for growth and cell division, bacteria and yeasts have such an effective potassium transport system that they can maintain high intracellular levels of this cation even when the latter is almost absent from the external fluid. However mutant strains of yeast and of *Escherichia coli* exist which are deficient in respiratory enzymes[14] and in their ability to accumulate potassium.[15,16] Such mutants require potassium in relatively high concentration in the medium for normal growth. In the mutant of *E. coli*, for example, growth fell by 50% when [K]$_o$ was only 2.5 mM, which caused [K]$_i$ to decrease to about 75 mM. In K-free medium cells lost 95% of their potassium and growth ceased completely but could be restored by the addition of potassium in high concentration. In spite of an intracellular concentration as low as 10 mM here, it was sufficient to sustain a slow synthesis of RNA, but there was a complete absence of protein synthesis. Inhibition of protein synthesis in K-deficient cells resembled that occurring in the presence of the antibiotic chloramphenicol, but in the latter case the amount of RNA in the cell actually increased while protein synthesis was inhibited.

The following steps have been recognized in the process of protein synthesis by the isolated microsomal fraction of tissues:

1. Formation of amino acid adenylate.
2. Formation of amino acyl-s-RNA.
3. Transfer of amino acid from soluble charged RNA to polypeptide chain in the ribosome.
4. The release of the completed polypeptide chain from the ribosomes.

Attempts have been made using the cell-free polyuridylic acid system[18] derived from liver to determine which of these steps needed potassium for activation. A tenfold increase in the rate of incorporation of amino acids into microsomes has been described in this system on the addition of potassium salts.[19] The incorporation of labeled phenylalanine into polypeptides using either the amino acid or its soluble RNA derivative was measured[20] in the presence of potassium and sodium. It was found that the transfer of amino acyl-S-RNA to polypeptide was 20 times faster in the presence of potassium than in the presence of sodium (Fig. 7.1). On the other hand, steps 1 and 2 appeared to be only slightly dependent on potassium. The rate of synthesis of phenylalanine-s-RNA was only twice as rapid in potassium as in sodium media. The third step however is a highly complex one requiring, in addition to charged soluble RNA and ribosomes, messenger RNA, polymerizing enzymes, guanosine triphosphate, and SH compound, and the ions Mg and K^+ or NH_4^+.[21]

Although these experiments strongly suggested that reaction 3 was the rate-limiting step in protein synthesis in low-K conditions, others[22] considered the release of the polypeptide chain from the ribosome, which must occur before another chain could be synthesized to be the potassium-dependent step. While this matter is still in doubt, investigations reported[23-25] indicate that potassium is required for at least one of the final stages of protein synthesis.

In fertilized sea urchins eggs,[26] a 16% reduction of intracellular potassium concentration from 155 to 130 mM blocked cell division and decreased the incorporation of ^{14}C-valine into proteins separated by electrophoresis on polyacrylamide gel. Phase contrast microscopy revealed that these cells did not then form a mitotic apparatus, the spindle and aster being either poorly developed or absent from the cell. The chromosomes

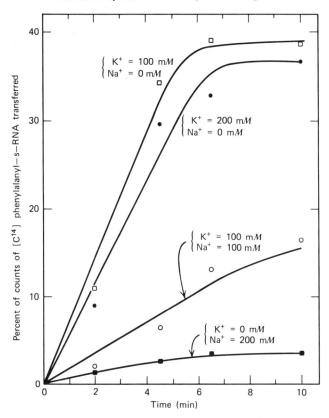

Figure 7.1 Kinetics of transfer of counts from ¹⁴C-phenylalanyl-s-RNA into polypeptide precipitable by trichloracetic acid, at various potassium and sodium concentrations. From M. Lubin & H. L. Ennis, *Biochim. Biophys. Acta* **80,** 614 (1964).

condensed into normal metaphase configuration, and although anaphase migration did not occur in the absence of the mitotic apparatus, they apparently underwent normal telophase disappearing into the nucleus, which must then have had a double complement of genetic material. It was concluded that protein required for the construction of the mitotic apparatus here could not be synthesized, a view supported by decreased amino acid incorporation.

POTASSIUM DEPRIVATION AND ENZYMATIC CHANGES IN MAMMALS

In higher animals the requirement of potassium for normal cell development and maintenance was also evident. When rats were fed a diet deficient in this cation, there was a sixfold increase in acid phosphatase activity in cells of the renal medulla[27] and a two- to fourfold increase in four other lysosomal enzymes.[28] Within a few days the cells accumulated dense granules containing cytochemically identifiable acid phosphatase activity. Much of the increase in lysosomes here was in the collecting duct cells of the renal papillae. Potassium deficiency also produced growth retardation and inability to concentrate urine.[29] In renal medulla there was a considerable increase both in cell volume and surprisingly in DNA content with increased synthesis of protein, RNA, and DNA.[30]

In mice,[31] potassium deprivation caused a 2- to 2.5-fold increase in kidney glucuronidase activity which increased linearly with time at least up to 40 days. Another tissue in which similar changes were seen was heart, but this may have been due to infiltration by leukocytes or fibroblasts. The absence of change in liver and similar organs may have been due to their ability to retain cellular K, even in the face of a fall in plasma level.

Associated with the inability to concentrate urine seen in K deprivation,[32] reductions were also found in the specific activities of mitochondrial enzymes in both homogenates and isolated mitochondria from red medulla of rats depleted of potassium.[33] It was suggested that the reductions might have impaired oxidative metabolism in the outer medulla where Henle's loop and collecting ducts responsible for water recovery from urine are located. In rats in which tissue K had been reduced from 119 ± 2.5 to 79.0 ± 5.8 m.mole·kg^{-1} tissue water with reciprocal changes in tissue sodium, the weight of the kidney had increased[29] by 34% and histochemical changes were evident.[34] The enzymes that showed a significant reduction in specific activity included oligomycin-sensitive ATPase, succinic dehydrogenase, and Mg-ATPase, but others, including microsomal (Na$^+$ + K$^+$)-ATPase and 5′-nucleotidase from medullary homogenate were not significantly changed in activity. It was not known whether the decrease in enzyme activity was a primary effect of potassium deficiency or a secondary one, arising from general metabolic disturbance.

While on the subject of the importance of mitochondrial function in the renal medulla and its apparent sensitivity to deprivation of potassium, it might be mentioned that mammalian erythrocytes that have lost their mitochondria and are dependent on anaerobic glycolysis for their energy can apparently function normally with relatively low intracellular potassium levels. In such LK erythrocytes the enzymes may have different thresholds for activation, because aerobic processes were depressed when only a third of cellular potassium was removed, while enzymes concerned with the earlier stages of glycolysis such as pyruvate kinase already discussed appeared to be maximally activated in the presence of as little as 11 mM K.[5]

THE ROLE OF POTASSIUM IN PURINE BIOSYNTHESIS

At the other extreme, potassium concentrations as great as 200 mM were needed for optimal activity in the enzyme system formyltetrafolate synthetase (THF-formylase) derived from plant sources. This enzyme catalyzes the following reactions leading to the formation of activated formate:

$$\text{THF} + \text{ATP} = \text{phosphoryl-THF} + \text{ADP} \tag{43}$$

$$\text{phosphoryl-THF} + \text{HCOOH} = \text{5-formyl-THF} + \text{P}_i \tag{44}$$

By transformylation, 10-formyl-THF is formed from 5-formyl-THF, and this is then used in a number of important synthetic reactions, including the synthesis of histidine from glutamic acid.

Activated formate in the form of N^5,N^{10}-anhydroformyl-THF is also required for transformylation steps in the synthesis of purines. In addition to the potassium requirement for the activation of formic acid, it is also apparently needed for the activation of enzymes concerned with two transformylation reactions in the synthesis of inosinic acid ribotide from glycinamide ribotide.[36] The first of these reactions catalyzed by glycinamide transformylase yields formylglycinamide ribotide:

$$O = C \underset{\diagdown NH-\text{ribose}-P}{\overset{\diagup CH_2-NH_2}{}} \xrightarrow[+ H_2O + K^+]{N^5 N^{10} \text{ anhydroformyl-THF}} O = C \underset{\diagdown NH-\text{ribose}-P}{\overset{\diagup CH_2-NH-CHO}{}} \tag{45}$$

At a later stage another transformylation takes place, this time to form a structure which yields inosinic acid by ring closure.[37] As inosinic acid is a precursor of both adenylic and guanylic acids; and as this final step is catalyzed by a K-requiring enzyme in pigeon and chicken livers, the significance of this cation in the pathways of purine biosynthesis in animals as well as in plants seems to have been established.

Finally a mutant strain, P-14, of *Aerobacter aerogenes* seemed to be unable to synthesize guanine from adenine or by pathways other than through xanthylic acid, and this required potassium in addition to NAD and an SH compound such as glutathione or cysteine.[39] The latter acid was then converted to guanylic acid by transamination from glutamine.

OTHER REACTIONS CATALYZED BY POTASSIUM AND THEIR CHARACTERIZATION

Many of the enzymes requiring potassium catalyze the transfer of phosphoryl groups to compounds having the general formula R—C(=X)OH. In the case of the phosphorylation of acetic acid by acetophosphotransferase, the X was the oxygen of a carboxyl group. In the case of pyruvic acid conversion to phosphoenolpyruvate by ATP, X was a carbon atom. A third type of compound phosphorylated in the presence of potassium and in which X may be represented by the group =N—R was exemplified by α-N-formylglycinamide ribonucleotide already mentioned.

Some have suggested[40] that the monovalent cation may help maintain a specific protein conformation needed for optimal catalytic efficiency. But others workers[41,42] more specifically proposed the formation of a tertiary complex between ATP, substrate, and cation during the phosphoryl transfer reaction. However not all enzymes requiring monovalent cation are concerned with such reactions. Aldolase, which catalyzes the aldolytic cleavage of fructose 1,6-diphosphate yielding dihydroxyacetone phosphate and the phosphate ester of glyceraldehyde, did not require metal ions in muscle and probably operated through a Schiff base in the enzyme–substrate complex.

In yeast and bacteria however, potassium ions were required for activation of this enzyme. The production of a carbanion intermediate with enol–keto tautomeric forms may have been enhanced by the cation.[43] A potassium requirement was also recognized in the case of some enzymes catalyzing peptide bond formation. An example is the formation of D-alanyl-D-alanine. Two distinct enzyme systems were responsible for the synthesis; one, requiring Mg^{2+}, gave rise to formation of an enzyme–AMP–D-alanine complex with liberation of pyrophosphate; the other, requiring both Mg^{2+} and K^+ for activation, formed D-alanyl-D-alanine and orthophosphate.[44] Potassium was also required for the synthesis of the tripeptide glutathione, which has been studied[45,46] in wheat germ extracts and proceeds in two steps, as follows:

glutamate + cysteine + ATP = γ-glutamylcysteine + ADP + P_i (46)

γ-glutamylcysteine + glycine + ATP = glutathione + ADP + P_i (47)

Enzymes catalyzing these reactions and requiring potassium have also been found in animal tissues.[47]

A number of important sulfhydryl compounds of the body seem either to be activated by potassium or to be associated with this cation in other respects. Methionine, the methylated derivative of homocysteine, is such a compound. In its activated form, as S-adenosylmethionine, it is an important methylating agent in the formation of choline, creatine, and epinephrine. The activation of methionine in rabbit liver and Bakers yeast[48] required the presence of potassium ions and involved the following general reaction:

$$l\text{-methionine} + ATP = S\text{-adenosylmethionine} + PP_i + P_i \quad (48)$$

This reaction was unique in that both pyrophosphate and orthophosphate were formed. It has been suggested that the S-adenosylmethionine synthetase mechanism, like other reactions requiring potassium, involves an enol–ketotautomer in a hypothetical intermediate.

These are but few of the many enzyme systems which appear to have a requirement for monovalent cations. The enzymes activated by potassium are also usually activated by Rb^+ and by NH_4^+ but show little or no activation by sodium or lithium. On the other hand, there are a few enzymes requiring the presence of sodium, and these are also activated by lithium but little or not at all by potassium, rubidium, and ammonium ions. In most cases the molar concentration for maximum activity of enzyme was about 10 mM.

The coupling of enzyme activation by monovalent cations with their active transport across membranes is a special case of the subject under review here and has already been discussed in Chapter 5, but it is worth noting that the two enzyme systems concerned, namely, the (Na^+ + K^+)-ATPase and the K-activated phosphatase systems in common with many nontransport enzymes activated by potassium are concerned with phosphoryl transfer reaction. The transport enzymes however are to be found in the membrane fraction of tissue homogenates rather than in the soluble fractions.

Finally a strong case has been made for the participation of potassium in enzyme reactions in which part of the substrate at least exhibits an enol–keto configuration, although it has been necessary in some cases to bolster up this view by the creation of hypothetical intermediates. The precise role of the monovalent cations in the catalytic process is as yet unknown, but their affinity for and complexation by molecules containing enol–keto tautomers may be relevant. For example, in the case of complexation of potassium ion by the ionophore valinomycin, the cation is surrounded by a three-dimensional ligand cavity bristling with oxygens of carbonyl groups. In the process of complex formation itself, the cation is apparently stripped of its water of hydration and the energy of interaction between oxygens and cation is sufficient to split hydrogen bonds within the valinomycin molecule, freezing more oxygen groups to chelate with the cation.

The macrotetralides nonactin, monactin, and diactin also have structures that enable them to surround the monovalent cation with an array of eight oxygen atoms in a quasicubic arrangement, but here oxygens alternate between carbonyl groups and furanose rings. Insofar as the interaction between monovalent cation and ionophore molecule exhibits a special affinity for the carbonyl group, and this group also seems to be found in many of the substrates of potassium requiring enzymes, it is possible that a study of such simpler complexes may provide information through structural and dynamic analysis which will uncover the mechanism of action of potassium in enzyme activation.

REFERENCES

1. A. F. Huxley, *J. Physiol.* **243**, 1 (1974).
2. O. Meyerhof, P. Ohlmeyer, W. Genther & H. Maier-Leibnitz, *Biochem. Z.* **298**, 396 (1938).
3. H. Lardy & J. A. Ziegler, *J. Biol. Chem.* **159**, 343 (1945).
4. O. Meyerhof & P. Oesper, *J. Biol. Chem.* **179**, 1371 (1949).
5. J. F. Kachmar & P. D. Boyer, *J. Biol. Chem.* **200**, 669 (1953).
6. E. J. Conway & P. T. Moore, *Biochem. J.* **57**, 523 (1954).
7. R. W. Hendler, *Nature* (Lond.) **207**, 1053 (1965).
8. G. Coleman, *Biochem. J.* **112**, 533 (1969).
9. G. Coleman, *Biochim. Biophys. Acta* **174**, 395 (1969).
10. J. D. Watson, *Bull. Soc. Chem. Biol.* **46**, 1399 (1964).
11. M. Molinaro & T. Hultin, *Exp. Cell Res.* **38**, 398 (1965).
12. T. Hultin, P. H. Näslund & M. D. Nilsson, *Exp. Cell Res.* **55**, 269 (1969).
13. T. Hultin, *Abhandl. Deut. Akad. Wiss. Berlin*, 379 (1968).
14. C. Reilly & F. Sherman, *Biochim. Biophys. Acta* **95**, 65 (1965).
15. C. Reilly, *Biochem. J.* **91**, 447 (1964).
16. H. L. Ennis & M. Lubin, *Biochim. Biophys. Acta* **50**, 399 (1961).
17. M. Lubin, Potassium and the Regulation of Protein Synthesis, in *The Cellular Function of Membrane Transport*, E. J. F. Hoffman, Ed., Prentice Hall, Englewood Cliffs, N.J., 1963.
18. M. W. Nirenberg & J. H. Matthaei, *Proc. Natl. Acad. Sci. U.S.* **47**, 1588 (1961).
19. H. Sachs, *J. Biol. Chem.* **228**, 23 (1957).
20. M. Lubin & H. L. Ennis, *Biochim. Biophys. Acta* **80**, 614 (1964).
21. T. Nakamoto, T. W. Conway, J. E. Allende, G. J. Spyrides & F. Lyman, *C. S. H. Sympos. Quant. Biol.* **28**, 227 (1963).

References

22. G. Webster & J. B. Lingrel, Protein Synthesis in Isolated Ribosomes in *Protein Biosynthesis*, R. J. C. Harris, Ed. Academic Press, New York, 1961, pp. 301–318.
23. D. Schlessinger, *Biochim. Biophys. Acta* **80**, 473 (1964).
24. M. Lubin, *Biochim. Biophys. Acta* **72**, 345 (1963).
25. J. B. Lingrel & G. Webster, *Biochim. Biophys. Acta* **61**, 942 (1962).
26. G. L. Meeker, *Exp. Cell Res.* **63**, 165 (1970).
27. A. B. Morrison & B. J. Panner, *Am. J. Pathol.* **45**, 295 (1964).
28. H. N. Aithal, F. G. Toback, S. Dube, G. S. Getz & B. H. Spargo, *Lab. Invest.* **36**, 107 (1977).
29. A. H. Manitius, H. Levitin, D. Beck & F. H. Epstein, *J. Clin. Invest.* **39**, 684 (1960).
30. A. B. Gustafson, L. Shear & G. J. Gabuzda, *J. Lab. Clin. Med.* **82**, 287 (1973).
31. C. E. Cleveland & R. T. Swank, *Biochem. J.* **170**, 249 (1978).
32. A. S. Relman & W. B. Schwartz, *Am. J. Med.* **24**, 764 (1958).
33. M. W. Weiner, L. A. Sauer, J. Torretti & F. H. Epstein, *Am. J. Physiol.* **221**, 613 (1971).
34. A. G. E. Pearse & C. R. Macpherson, *J. Pathol. Bacteriol.* **75**, 69 (1958).
35. J. V. Evans & A. T. Phillipson, *J. Physiol.* **139**, 87 (1957).
36. J. C. Rabinowitz, Folic Acid in *The Enzymes*, P. P. Boyer, H. Lardy & K. Myrback, Eds., Academic Press, London, Vol. 2, 1960.
37. J. G. Flaks, M. J. Erwin & J. M. Buchanan, *J. Biol. Chem.* **229**, 603 (1957).
38. E. F. Gale & J. P. Folkes, *Biochem. J.* **53**, 483 (1953).
39. B. Magasanik, H. S. Moyed & L. B. Gehring, *J. Biol. Chem.* **226**, 339 (1957).
40. H. J. Evans & G. J. Sorger, *Ann. Rev. Plant Physiol.* **17**, 47 (1966).
41. J. B. Melchior, *Biochemistry* **4**, 1518 (1965).
42. J. M. Lowenstein, *Biochem. J.* **75**, 269 (1960).
43. C. H. Suelter, *Science* (N.Y.) **168**, 789 (1970).
44. F. C. Neuhaus, *J. Biol. Chem.* **237**, 778 (1962).
45. G. C. Webster & J. E. Varner, *Arch. Biochem. Biophys.* **52**, 22 (1954).
46. G. C. Webster & J. E. Varner, *Arch. Biochem. Biophys.* **55**, 95 (1955).
47. E. J. Snoke & K. Bloch, *J. Biol. Chem.* **199**, 407 (1952).
48. C. L. Woronick & M. J. Johnson, *J. Biol. Chem.* **235**, 9 (1960).
49. R. E. Parks Jr., E. Ben-Gershom & H. A. Lardy, *J. Biol. Chem.* **227**, 231 (1957).
50. V. Paetkau & H. A. Lardy, *J. Biol. Chem.* **242**, 2035 (1967).
51. O. H. Lowry & J. V. Passonneau, *J. Biol. Chem.* **241**, 2268 (1966).
52. J. A. Muntz & J. Hurwitz, *Arch. Biochem. Biophys.* **32**, 137 (1951).
53. F. C. Happold & R. B. Beechey, *Biochem. Soc. Symp.* **15**, 52 (1958).
54. J. S. Nishimura & D. M. Greenberg, *J. Biol. Chem.* **236**, 2684 (1961).

55. S. H. Mudd & G. L. Cantoni, *J. Biol. Chem.* **231,** 481 (1958).
56. C. Milstein & A. O. M. Stoppani, *Biochim. Biophys. Acta* **28,** 218 (1958).
57. J. R. Stern & C. S. Hegre, *Fed. Proc.* **26,** 605 (1967).
58. R. Rodnight and B. E. Lavin, *Biochem. J.* **93,** 84 (1964).
59. H. Bader and A. K. Sen, *Biochim. Biophys. Acta* **118,** 116 (1966).
60. H. Yoshida, F. Izumi & K. Nagai, *Biochim. Biophys. Acta* **120,** 183 (1966).
61. J. R. Bertino, *Biochem. Biophys. Acta* **58,** 377 (1962).
62. R. W. Holley, E. F. Brunngraber, F. Saad & H. H. Williams, *J. Biol. Chem.* **236,** 197 (1961).
63. R. S. Schweet, R. Arlinghaus, J. Schaeffer & A. Williamson, *Medicine (Baltimore)* **43,** 731 (1964).
64. A. J. Giorgio & G. W. E. Plaut, *Biochim. Biophys. Acta* **139,** 487 (1967).
65. J. B. Edwards & D. B. Keech, *Biochim. Biophys. Acta* **159,** 167 (1968).
66. M. Marshall, R. L. Metzenberg & P. P. Cohen. *J. Biol. Chem.* **236,** 2229 (1961).
67. P. Datta and L. Prakash, *J. Biol. Chem.* **241,** 5827 (1966).
68. R. W. Von Korff, *J. Biol. Chem.* **203,** 265 (1953).
69. L. T. Webster Jr., *J. Biol. Chem.* **241,** 5504 (1966).

8
Active Transport of Potassium in Epithelial Tissues

A usual starting point for a consideration of epithelial transport of ions has been the classical experiments of Ussing and his colleagues with frog skins.[1,2] When these were interposed between two chambers filled with Ringer fluid and the electromotive force (E.M.F.) between the chambers measured, it was found that the inner surface (chorion or serosal) of the skin was positively charged by up to 100 mV with respect to the outer (epithelial or apical) surface. When skins were short-circuited by the application of an opposing voltage from a battery and the current in the external circuit measured, it was found that this short-circuit current of up to 80 μA·cm^{-2} was proportional to the net flux of sodium ions from the outside to the inside of the skin. Net sodium flux was usually measured by labeling the fluid bathing the inside of the skin with ^{22}Na and that at the outer surface with ^{24}Na. When the difference between these fluxes was converted to electrical units of charge, the magnitude of the resulting current accounted for that measured directly. If the sodium concentration of the fluid at the apical surface, $[Na]_a$, was reduced, net transport and current could still be measured down to 1 m.mole·l^{-1} when the inner surface was bathed with Ringer.

KOEFOED-JOHNSEN USSING MODEL OF TRANSEPITHELIAL TRANSPORT

It was suggested that sodium ions leaked into the epithelial cells from the external fluid and were pumped out at the serosal side of the epithelial cell layer by a sodium–potassium linked pump operating in an electrically neutral manner. The leakage of potassium out of the cells at this pumping membrane was believed to contribute to the transepithelial current. According to this view[3] the outer surface of the skin was freely permeable to sodium but not to potassium, while the reverse was true of the inner surface of the epithelial layer so that the skin potential was the sum of two diffusion potentials, one for sodium and the other for potassium:

$$\text{E.M.F.} = \frac{RT}{F} \ln \frac{[\text{Na}]_a}{[\text{Na}]_i} + \frac{RT}{F} \ln \frac{[\text{K}]_i}{[\text{K}]_s} \qquad (49)$$

where $[\text{Na}]_a$ and $[\text{Na}]_i$ are sodium concentrations in external fluid and intracellular fluids, respectively, and $[\text{K}]_i$ and $[\text{K}]_s$ are potassium concentrations in intracellular fluid and in fluid bathing the serosal side, respectively. One would expect from this model that two voltage steps would be encountered on passing a microelectrode through the epithelial cell layer. Unfortunately amphibian skin was not easily penetrated by microelectrodes. But when blisters were produced on skin by various techniques,[4,5] areas of up to 1 cm² of epithelium could be separated from the corium, the thick innermost layer of loose connective tissue, to facilitate penetration by the electrode. Removal of the corium with the basement membrane left exposed the epithelial cells of the stratum germinativum which still possessed the transport characteristics of the intact skin. When microelectrodes were advanced from the inner surface, a series of distinct stable potential steps were found making up the electrical profile.[4]

Flux measurements made in open circuit have shown chloride to be passively distributed through the skin, and this ion may enter the cells as a counter ion for sodium transported.[1] Replacement of external chloride by impermeant anions such as methylsulfate increased the skin potential, indicating the shunting effect of chloride movement on this potential. But this did not happen in toad bladder.[5]

The toad bladder, which has also been widely used to study epithelial transport, acts functionally as an extension of the kidney, being similar to the distal convoluted tubule and collecting duct of the mammalian nephron. The cell layer of interest for transport in this tissue is the mucosal epithelial cell layer, which contains four cell types.[6] Those concerned with transport are the granular cells, which occupy most of the luminal surface, and the mitochondria-rich cells, which comprise 5–10% of the total. These cells span the gap between luminal surface and basement membrane. In this cell layer, sodium enters at the luminal surface and is pumped out at the serosal surface. Two voltage steps have been found when the bladder was penetrated by microelectrodes from the luminal or mucosal surface: the inside granular cells being positive with respect to the urine, and a further positive step found when the microelectrode passed through the serosal membrane.

Although the finding of two voltage steps on penetrating the frog skin and toad bladder with microelectrodes was consistent with the Koefoed–Johnson Ussing model of the epithelial potential being the sum of two diffusion potentials in series, eq. (49), the potential could also have been due in part at least to the electrogenic pumping of sodium ions across the serosal membrane[7] in the absence of tight coupling of Na–K exchange.

The characterization of potassium movement across the epithelial membranes presents a number of problems. First it should be mentioned that there is strong evidence that the apical luminal membrane has a low potassium permeability. For example,[8] in the case of the toad urinary bladder, only 1.5% of cellular potassium equilibrated with ^{42}K in the lumen over a period of 60 min. It is important to know the stable potentials across luminal and serosal membranes and the activity of potassium ions within the cells as well of membrane conductances or unidirectional fluxes and their modification if any by ouabain and other substances and conditions which affected short-circuit current and transepithelial transport in order to fully understand the mechanism of potassium transport in epithelial tissues.[9]

Let us first consider the ouabain-sensitive sodium pump which is situated at the basolateral or serosal membrane. Its location has been confirmed by autoradiography using ^3H-ouabain.[10] The observation[9] that ouabain inhibited transepithelial sodium transport when present in the serosal medium but not when present in luminal fluid also confirmed the location of the pump. The relationship between the short-circuit current and ouabain binding has been studied in the Ussing chamber.[10] All connective tissue was removed from the frog skin by treatment with collagenase, and as an extracellular marker ^{14}C-mannitol was used to correct for ouabain in this space. It was found that the binding of ^3H-ouabain by such split skins and the inhibition of the short-circuit current (I_{SC}) proceeded as a hyperbolic function of time. When the number of ouabain molecules bound was plotted against the percent inhibition of I_{SC}, a straight line was obtained, but on average no inhibition was evident until about a third of the binding had taken place. It was then found that skins could be separated into two groups, those with a high I_{SC} in which inhibition occurred with very little binding and others with a smaller I_{SC} in which half the sites had to be occupied by ouabain before any fall in I_{SC} was evident. It was suggested that in the latter skins, inhibition of some sites caused recruitment of others, thus maintaining the pumping of sodium at a constant level. These findings support earlier evidence that ouabain binding at the basolateral membrane was correlated with inhibition of I_{SC}.

MEASUREMENT OF INTRACELLULAR POTASSIUM ACTIVITY AND RELATIVE ION PERMEABILITIES

Intracellular potassium activity was measured[11] in the epithelial cells of the urinary bladder of rabbit by means of ion-selective microelectrodes. The activity was found to have a value of 72 m.mole·l^{-1} cell water, decreasing to 32 m.mole·l^{-1} after 1 hr of exposure to 10^{-4} M ouabain. Intracellular chloride activity was similarly measured and found to be 15.8 m.mole·l^{-1}. By indirect methods intracellular sodium activity in the steady state was estimated to be less than 13 m.mole·l^{-1}. The membrane potential across the serosal membrane was -53.8 mV, and this decreased to -26 mV in the presence of ouabain.

In the presence of sodium chloride, the cell interior was found to be always negative with respect to the serosal solution and independent of the rate of sodium transport. Although chloride ions appeared to be at electrochemical equilibrium across the serosal membrane, potassium ions were not, and therefore needed to be actively transported into the cells from the serosal fluid.

The ratios P_{Na}/P_K and P_{Cl}/P_K were determined by use of the Goldmann equation after the potential of the serosal membrane had been measured under conditions where $[K]_0$ was varied, namely, keeping the KCl product constant, and with constant $[Cl]_0$ and equimolar replacement of Na_0 by K_0 over the $[K]_0$ range of 1.9 to 15 mM. These ratios were found to be 0.044 and 1.17, respectively. Using these permeability coefficients and measured intracellular concentrations, it was feasible to calculate the diffusion potential under steady-state conditions for comparison with measured potentials. In this way the possibility of electrogenic pumping of sodium could be explored.

ELECTROGENIC OR NEUTRAL PUMP

Two methods were used to determine whether the pump was electrogenic or tightly coupled, exchanging Na and K, ion for ion. It had been found[12] that the polyene antibiotic nystatin, on addition to mucosal solution, produced a 100-fold increase in the conductivity of the apical membrane while the serosal membrane hyperpolarized by about 10 mV, with alteration of its resistance. This could have been due to sodium leak at the apical membrane, the resulting rise in [Na]$_i$ stimulating a sodium pump with a Na:K coupling ratio greater than unity. The other possibility was that the pump stimulated by sodium entrance at the apical membrane had a 1:1 coupling ratio but that increase in E_K resulting from K influx caused the serosal membrane to hyperpolarize. In order to distinguish between these two possibilities, the following procedure was adopted.

The sodium chloride of the mucosal Ringer was replaced by potassium sulfate and 60 μl nystatin was added. When a steady state had been reached, sodium-containing serosal Ringer was also replaced by potassium sulfate. This caused the transepithelial potential to fall to zero. Equal amounts of NaCl were then added to both bathing fluids to final concentrations of 13.3, 26.7, 40, 53.3, and 66.7 mM. This produced a potential across the serosal membrane of about −20 mV, which was reduced by 80% within 3 min of adding $10^{-4}M$ ouabain to the serosal solution. The ouabain also inhibited the I_{SC}. This electrogenic sodium pumping was not detectable at normal pumping rates, so the serosal membrane potential under those conditions was basically a diffusion potential with potassium and chloride the main contributors. Addition of amiloride, which blocked sodium influx at the mucosal membrane, or the replacement of sodium in both mucosal and serosal fluids with choline did not alter significantly the serosal membrane potential, indicating that the direct contribution of electrogenic sodium pumping to membrane potential must be small under normal conditions.

The uptake of potassium into epithelial cells in rabbit urinary bladder is likely to be accomplished through the activity of a ouabain-sensitive (Na$^+$ + K$^+$)-ATPase transport system at the serosal membrane and regulated through variation in its coupling ratio. Its passive uptake in response to membrane hyperpolarization by sodium extrusion would seem to be unlikely in this tissue. There were conditions in other epithelial tissues in which the inhibition of transepithelial transport did not result in loss of potassium from the cells. For example, direct measurement of cellular potassium content of toad bladder epithelium have indicated no change when transepithelial transport of sodium was blocked by removing sodium from the luminal medium or by adding 10^{-4}–10^{-3} M amiloride.[13] Also when two sets of hemibladders were incubated, one set under open circuit conditions and the other in closed circuit,[14] there was no significant difference in either the cellular potassium concentration or water content of the cells, suggesting the independence of the maintenance of intracellular potassium concentration from transepithelial sodium transport.

In the urinary bladder of toad a_K^i has also been measured by ion-selective microelectrodes under open and closed circuit conditions in a number of laboratories.[15-17] Values ranging from 39.3 m.mole to 43.0 m.mole·l^{-1} cell fluid were found which are considerably less than those in the rabbit. The addition of ouabain or the omission of potassium from the bathing fluid produced a loss of potassium from the epithelial cells reflected in a fall of a_K^i, and this was accompanied by a decrease in I_{SC}.

Table 8.1 shows values of transepithelial potential difference (PD) and membrane potentials across serosal and mucosal membranes, E_m^s and E_m^m, along with short-circuit current (I_{SC}) and calculated potassium equilibrium potentials (E_K) across the serosal membrane, under various conditions. From these data several important conclusions may be drawn. First, the fact that the internal negativity at the serosal membrane was less than E_K indicates the necessity for potassium ions to be actively pumped into the cells across this membrane. Although the inhibitor ethacrynic acid[15] reduced the transepithelial potential by about 90% and had a similar effect on I_{SC}, a_K^i was virtually unchanged. This diuretic has been found to block an ouabain-insensitive sodium transport in kidney which was highly temperature dependent, stimulated by angiotensin, and probably electrogenic. A distinction has been made between this pump[18] and the Na–K exchange pump, which is relatively insensitive to temperature, refractory to diuretics, and probably responsible for maintaining the normal intracellular levels of potassium and sodium.

It might be concluded from the result with toad bladder that the maintenance of intracellular potassium in the epithelial cells is not dependent on transepithelial sodium transport but that the converse is true, namely, that sodium transit through the cells depends on the maintenance of normal a_K^i. This view is supported by the kinetics of recovery of a_K^i and of I_{SC} from the effects of addition of ouabain or removal of potassium from the medium. For example,[17] addition of 10^{-2} M ouabain to the serosal medium reduced a_K^i by 56–67% and I_{SC} by 96–100%, while 5×10^{-4} M ouabain reduced these by 40–55% and 63–68%, respectively. On washing out the ouabain in the latter case, a_K^i was restored within 2 hr (Fig. 8.1), while I_{SC} was still only about 50% of normal after more than 3 hr. As expected, the removal of potassium from the bathing fluid reduced a_K^i and I_{SC}, the former by up to 80% and the latter by up to 76%. On restoring external potassium to its normal level, its intracellular activity was restored within 1 hr, while I_{SC} had shown no significant increase within this time and had not increased to more than 50% of its control level 1 hr later. It was concluded that the process of potassium accumulation and of transepithelial sodium transport were separate transport processes at the basolateral membrane of the cells.

Table 8.1 Effects of Some Transport Inhibitors on the Intracellular Potassium Activity a_K^i, Short-Circuit Current I_{SC}, and Membrane Potentials in the Toad Bladder Epithelium[a]

Conditions	N	PD (mV)	I_{SC} (μA)	E_m^s (mV)	E_m^m (mV)	E_K (mV)	a_K^i (m.mole·l^{-1})
Control	76	25.1 ± 7.5	49.3 ± 12.3	−12.7 ± 5.0	12.4 ± 3.4	−67.2 ± 0.5	39.3 ± 0.7
Ouabain, 10^{-3} M 3 hr, both sides	10	−2.5	−3.5	0.4 ± 0.3	−2.1 ± 0.3	−28.8 ± 0.5	8.1 ± 0.3
Ethacrynic acid, 10^{-3} M, serosal side, 3 hr	16	0.9	7.6	−0.4 ± 0.1	0.5 ± 0.1	−67.1 ± 0.9	36.8 ± 1.4
Rotenone, 10^{-5} M 1 hr, both sides	11	1.3	1.2	−0.6 ± 0.1	0.7 ± 0.1	−53.5 ± 5.3	26.4 ± 2.2
K-Free, 3 hr, both sides	12	2.7	6.5	−1.2 ± 0.1	1.5 ± 0.1	−85	11.1 ± 1.2

[a] N = Number of PD measured. All potentials given as serosal side positive.[16] PD is transepithelial potential difference; E_m^s and E_m^m, membrane potentials at serosal and mucosal borders, respectively.

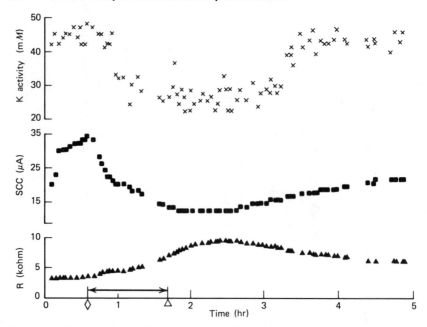

Figure 8.1 Time course of effects associated with ouabain on a_K^i, I_{SC}, and transepithelial electrical resistance R. The tissue was bathed with a serosal Ringer containing 5×10^{-4} M ouabain during the period indicated by solid arrow. From J. DeLong & M. Civan, *J. Membr. Biol.* **42,** 19 (1978).

Support for the Koefoed–Johnsen Ussing model[3] of epithelial E.M.F. seemed to come from experiments in which this potential was attenuated by progressive substitution of potassium for sodium in the external fluid or the luminal surface of toad bladder and where there was found to be a linear relationship between the E.M.F. and log of the sodium concentration in this medium. However it was subsequently found[19] that increase in potassium concentration at the serosal membrane depolarized the mucosal membrane also, showing an interdependence between them. Furthermore replacement of all the serosal sodium by potassium resulted in a potential across this membrane which could not be accounted for by passive diffusion of ions. Finally it should be mentioned that three substances whose action was believed to be confined to only one border of the epithelial layer, namely, vasopressin, ouabain, and amiloride, have been found to influence both membranes. For example, ouabain seemed to inhibit sodium entry into cells of urinary bladder from the lumen when the sodium concentration at that border was less than 20 mM.[20]

The measurement of fluxes of radioisotopes of potassium in epithelial tissue has raised questions about its state in the cells and of the possibility of its compartmentalization. In toad urinary bladder, for example, the kinetics of exchange of labeled for unlabeled potassium indicated at least two pools.[7] One making up a quarter to a third of cellular potassium exchanged with a half-time of about 30 min, while the remainder exchanges at a much slower rate. Even after 19 hr of incubation at 18–20°C, all the cellular potassium had not exchanged with ^{42}K. The amount of potassium in the first pool was similar to the amount that exchanged with Rb$^+$ introduced into the medium and to the amount of potassium lost from the cells after up to 2 hr of incubation with K-free Ringer or in the presence of ouabain. This has led to the suggestion that only about a third of cell potassium here is in free solution in cytoplasm. In other studies[21] with toad bladder it was found that only about 10% of cell K had exchanged within 1 hr, and only this fast-exchanging fraction changed when transepithelial transport was modified. As mentioned already, there are four different cell types in urinary toad bladder, so perhaps it would be more surprising to find one rate constant for potassium flux. Also the K fluxes at the serosal border were much greater than at the apical border, as might have been expected from the higher permeability of the latter. In view of the apparent increased exchangeability of potassium in tissues respiring vigorously compared with those kept in anaerobic conditions,[22] mitochondrial activity may participate in the process.

Ouabain, which inhibited sodium exit at the serosal border thereby reducing transepithelial transport by 80%, had no effect apparently on rate constants for potassium fluxes nor on pool sizes in bladders mounted in normal Ringer.[23] In the presence of this inhibitor however the removal of potassium from the serosal fluid resulted in a significant decrease in the rate constant for its efflux into this medium, indicating the presence of a K–K exchange process. Since such findings were difficult to reconcile with the suggested tight coupling of Na–K transport across the serosal membrane, it seemed likely that in the presence of ouabain the Na–K exchange might be replaced by K–K exchange, due to reduced affinity of internal pump sites for sodium relative to potassium. The latter exchange required the presence of sodium in the mucosal fluid and was inhibited by furosemide,[24] which appeared to inhibit transepithelial transport in kidney.

TRANSEPITHELIAL POTASSIUM TRANSPORT

This is a subject which has been researched to a lesser extent than epithelial sodium transport, first because there are fewer tissues in which it is present, and second because these tissues are generally less accessible to investigation.

One of the earliest examples studied in any detail was that which occurred in the midgut of the larval state of the silkworm *Cecropia*,[25] which lives on a potassium-rich diet. The midgut tissue has available sources of energy to prevent the entrance of potassium into the blood while nutrients are absorbed from the lumen. The luminal surface appeared to have a low permeability to this cation, which was pumped from the serosal side to the lumen. The E.M.F. across the gut wall varied from about 68 mV (serosal border negative) in the early larval stage to a maximum of 124 mV shortly after evacuation of the gut as a preliminary to the pupal stage.[26] Within 12 hr of attainment of this maximum, the potential fell precipitously to zero and did not increase again.

During the stage when the E.M.F. was about 84 mV the removal of potassium from the serosal border of the gut caused the potential to fall to zero, and about 83% of the 400 $\mu A \cdot cm^{-2} I_{SC}$ could be accounted for by the net flux of potassium toward the lumen.[27] Subsequently with improved techniques[28] it was found that all of the current was carried by potassium. The current was inhibited by anoxia and DNP but was relatively insensitive to ouabain.[29] The measurement of unidirectional K fluxes revealed that the net flux toward the lumen was about 32 times greater than predicted for passive diffusion.[27] Rubidium, which closely resembles potassium chemically, could substitute for the latter in maintaining the transepithelial potential.[30]

POTASSIUM TRANSPORT IN THE COCHLEA

It was known for a long time that the endolymph contained in the scala media (Fig. 8.2) of the cochlea was a self-contained isolated system independent of perilymph of the scala vestibuli and scala tympani, which is connected to the cerebrospinal fluid. It was then discovered that the endolymph differed from the latter in being high in potassium and low in sodium like cytoplasm.[31,32] Values of 150 mM have been quoted for potassium and 2 mM for sodium,[33] and these values are maintained by cells of the stria vascularis which line the outer wall of the scala media.

Another unusual feature of the cochlea is that when microelectrodes were introduced into the scalae, it was found that the scala media containing the endolymph was positively charged by about 80 mV with respect to the scalae containing perilymph. This potential was extremely sensitive to oxygen lack, falling to zero when ventilation of an anesthetized guinea pig was interrupted for about a minute and even becoming

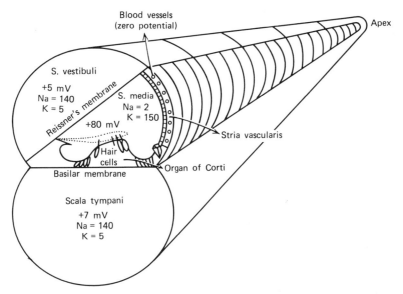

Figure 8.2 Schematic diagram of the cochlea. From Johnstone & Sellick, Q. Rev. Biophys. **5**, 1 (1972).

negative over a longer period. Attempts have been made to relate this potential to diffusion of ions down electrochemical gradients. More recently the ionic composition of the cochlear fluids has been determined under ideal conditions by means of ion-selective microelectrodes,[34,35] and this has helped in the calculation of equilibrium potentials of the principal ions present. The equilibrium potentials in cochlear endolymph of potassium, sodium, and chloride in relation to perilymph were calculated to be −86, 120, and 0 mV, respectively, so none of these ions was at equilibrium in its distribution between the two fluids. However evidence has been provided[36] to show that, under anoxic conditions when the endolymph has become negatively charged with respect to the perilymph, this is generated by the flux of potassium from the former to the latter fluid.

What then can be said of the origin of the endocochlear potential under normal conditions? It could be due to (a) diffusion of sodium into the endolymph,[33] (b) pumping of chloride in the opposite direction, or (c) electrogenic pumping of potassium into the endolymph. When the concentration of sodium in the endolymph was increased from its normal value of 2 mM to 20 mM by perfusion of the scala media, thereby decreasing the electrochemical potential for sodium from about 120 mV to

59 mV, instead of decreasing, the endocochlear potential (EP) actually increased in magnitude, so sodium diffusion was probably not responsible for its generation. Furthermore when Reissner's membrane was destroyed so as to cause mixing of cochlear fluids and elimination of the [Na] gradient, a large EP could still be detected by a microelectrode placed near the stria vascularis. When the chloride in the endolymph was reduced from about 110 mM to less than 3 mM by perfusion, there was no change in the magnitude of the EP. It thus seems unlikely that electrogenic pumping of this anion from the endolymph produced the potential. As none of the ions under consideration was at equilibrium, it must be assumed that sodium was pumped out of the endolymph along with chloride, while potassium ions were pumped inward. It has been suggested that there may be two cation pumps in the stria vascularis, an electrogenic potassium pump on the border facing the endolymph, responsible for the EP, and a Na–K pump on the serosal border of the epithelial cells. The latter has been envisaged as normally inactive because of the low concentration of sodium ions in the scala media. However it is significant that the EP is rapidly eliminated by ouabain[38,39] as well as by cyanide,[35] so the Na–K pump may function continually.

The idea that two pumps were present came from experiments with the diuretic ethacrynic acid. When it was injected intravenously into guinea pigs[34] at a concentration of 20–45 mg·kg^{-1} body weight, it reversibly reduced EP while increasing the sodium concentration of the endolymph from 1 mM to 5–8 mM at 20 min after injection. At higher concentrations, EP was reduced to negative values[35] which were rapidly suppressed to zero by anoxia. It has been suggested[35] that the negative potential might be due to electrogenic pumping of sodium toward the blood, generating about -18 mV through a cation pump whose coupling ratio was 3Na:2K. The potential found after application of ethacrynic acid was -13 mV. Although 10^{-3} M ouabain introduced into perilymphatic perfusate also changed EP to a negative value, albeit more slowly than the ethacrynic acid, the resulting potential was not sensitive to anoxia or to metabolic inhibitors and was probably due to the diffusion of potassium ions from the endolymph to perilymph. That such diffusion is normally present was suggested by an increase in EP seen when the scala vestibuli was perfused with high-K Ringer.[39]

Because of the small size of the cochlea and its relative inaccessibility, attempts to measure[40,41] the unidirectional fluxes of ^{42}K and ^{22}Na or ^{24}Na between scalae have been generally unsuccessful because of contamination of fluid samples. In one case ^{24}Na and ^{42}K were injected into the perilymph and sampled in endolymph. Instead of using the micropuncture technique, the cochlea was frozen with liquid nitrogen and dissected out in this state. The perilymph and endolymph were separated on the basis of their different melting points. In the other case[41] the isotopes ^{22}Na and ^{42}K were injected into the blood of cats and sampled in endolymph removed by micropuncture.

The location of the cation pumps, whether at the luminal or serosal border of the epithelial cell layer, has not yet been finally settled. It is intriguing to note that only one potential step has been detected by microelectrodes traversing the stria vascularis.[42] The experiment in which application of ethacrynic acid either through the blood or through perilymphatic perfusate converted a positive EP into a negative EP which either decayed spontaneously to zero over a period of 80 min or could be rapidly reduced by anoxia would seem to indicate the presence of two pumps, one of which was probably the ouabain-sensitive (Na$^+$ + K$^+$)-ATPase. Indeed high activity of this enzyme has been detected[43] in preparations from the stria vascularis. But it might also be added that this tissue in common with other potassium transporting tissues[28] also has an oxygen consumption higher than that of kidney and brain.[44]

It seems unlikely that the ouabain-sensitive Na–K pump would be at the endolymphatic border, as it would pump sodium into the cell and another sodium pump would be required at the serosal border to rid the cell of this cation. The response of the EP to ouabain added to the perilymph but not to endolymph is also consistent with this location. Although the paucity of information on membrane potentials at the borders of the cells makes it impossible to localize the pumps with certainty and to decide on the presence or absence of a second pump, a consideration of such measurements[45] in the other potassium-transporting epithelium, that of Cecropia midgut, might be helpful in this respect. When microelectrodes were advanced from the blood side of the midgut, it encountered three small negative steps of -3, -10, and -25 mV, respectively, followed by a jump to $+105$ mV (Fig. 8.3), which was the midgut potential. When the electrode was advanced from the luminal side the profile was reversed, but sometimes the negative potential was "missed" and only the one large positive potential was seen, as was the case in the cochlea. The maximum negativity encountered was 27 mV, and the average positive potential for 55 profiles was 125 ± 3.5 mV, the largest value recorded being $+187$ mV.

When the midgut was made anoxic by bubbling with nitrogen, the negative potential was unaffected, but the large positive potential was reduced more or less in line with the fall in midgut potential. The negative potential on the other hand was changed by substituting sodium for potassium in the fluid at the serosal border in a manner expected of a potassium diffusion potential, with a slight leakage of some other ions.

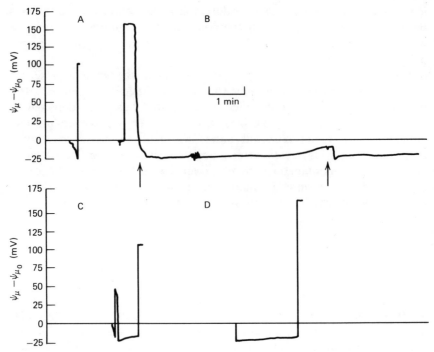

Figure 8.3 Potential profile recorded when midgut in oxygenated solution was impaled (A) from blood side, showing three small negative plateaus and a single large positive step to full midgut potential, (B) from lumen, missing negative plateaus and recording full midgut potential. The microelectrode was then retracted slowly (arrow) and a negative plateau recorded for 5 min. (C) Profile showing a rarely observed transient smaller positive potential superimposed on an otherwise "typical" profile. (D) Profile showing sustained negative plateau followed by the largest positive step recorded (+187 mV). From J. L. Wood, P. S. Farrand & W. R. Harvey, *J. Exp. Biol.* **50**, 169 (1969).

This negative potential had its equivalent in the anoxic cochlea.[36] The anoxia-sensitive negative potential of cochlea uncovered by ethacrynic acid might on the other hand have been due to an electrogenic Na–K pump at the serosal membrane.

It was concluded that the profile in the midgut was made up of two potentials in series—a potassium diffusion potential at the serosal surface and an electrogenic potassium pump at the luminal border, pumping this cation into the lumen and being responsible for the positivity of the fluid at this border. If a comparison between the midgut and stria vascularis is valid, then one would expect that an electrogenic potassium pump at the cell border in contact with endolymph would be responsible both for the positive EP and the maintenance of the high potassium concentration in this fluid.

POTASSIUM TRANSPORT IN SALIVARY GLANDS

In general, mammalian salivary glands secrete potassium ions from blood to lumen.[46] The primary secretion of saliva occurs in the acinar cells and is usually an intermittent process under autonomic control. Stimulation of cholinergic and α-adrenergic receptors increases membrane permeability to potassium, with consequent hyperpolarization of the cells.[47] The most recent estimates of membrane potential across the serosal membrane of acinar cells have been[48] in the region of -70 to -80 mV, which are significantly higher than earlier values. The potassium equilibrium potential based on measured intracellular potassium concentration of 150 m.mole·l^{-1} cell fluid was -90 mV. This would suggest that potassium ions must be pumped into the cells. Primary saliva resembles serum in its composition, but it undergoes modification first in response to stimulation of secretion and second as a result of ion transport in cells lining the salivary duct.

When ^{86}Rb release was used as an index of K efflux,[49] the cholimimetic ester carbachol induced a biphasic increase in this process in slices from rat sublingual gland. The first phase was transient, lasting 2–4 min, and was followed by a sustained, slowly decreasing phase that was dependent on the presence of calcium in the bathing fluid. The addition of the divalent ionophore A23187 also promoted the slower phase of ^{86}Rb release, apparently by inducing an entrance of calcium ions into the cells. In lacrimal acinar cells of the mouse,[50] microinjection of Ca^{2+} also caused membrane hyperpolarization and a fall in membrane resistance which was due apparently to an increase in potassium conductance. Carbachol also reduced the potassium content of the tissue *in vitro* by 30%.

It has been possible to study potassium movements in salivary glands of anesthetized cats and dogs through the use of external potassium-selective microelectrodes.[51] It was found by this means that external potassium concentration increased from 2.3 m.mole·l^{-1} to 15.2 m.mole·l^{-1} in dogs and from 2.2 m.mole·l^{-1} to 18.7 m.mole·l^{-1} in cats following electrical stimulation of the parasympathetic chorda lingual nerve. However the concentration fell once more at the end of stimulation, probably because of active uptake by the gland cells. There appeared to be a parallel between the amount of potassium lost from the cells and the rate of secretion by the gland.[52] The link between the two processes has yet to be defined. It was suggested that during potassium release from the acinar cells an equivalent amount of sodium entered the cells and that this was the trigger for secretion. It is not clear how a net influx of sodium may be reconciled with the membrane hyperpolarization induced by stimulation.

When Ussing's flux equation was applied[53] to the unidirectional potassium fluxes across the cells lining the salivary duct of the rat, the observed flux ratio was greatly in excess of the predicted value for passive movement, supporting earlier contentions that potassium release into the duct lumen and movement of sodium in the opposite direction were active processes. Variation of potassium concentration within the lumen did not appear to change the transductal potential difference. The serosal membrane was found[54] to contain a ouabain-sensitive ($Na^+ + K^+$)-ATPase responsible apparently for the pumping of sodium into the blood and of potassium in the opposite direction. Unfortunately there is little information at present about the membrane potential at the luminal surface of the epithelial cells in the salivary gland. Its apparent similarity to that of the serosal surface[55] in unstimulated gland may be due to large shunt pathways for current flow through the tight junctions between cells.

In order to determine whether sodium entrance and potassium loss from acinar cells *per se* was responsible for initiation of secretion, the former changes were brought about by means other than nerve stimulation, namely by exposure of the tissue to low temperature or by removal of external potassium.[52] On returning to normal conditions, a net uptake of potassium and active extrusion of sodium took place from the cells as happened after parasympathetic stimulation or addition of ACh. However only in the latter cases was salivary secretion produced, so that the ionic movements were probably secondary to or independent of the formation of primary saliva. Although the potassium movement into the lumen is generally considered to be passive, when its concentration in the primary saliva rises to 160 mM after sympathetic stimulation it is difficult to be convinced that active secretion is not taking place.

In vivo and *in vitro* microperfusion and micropuncture techniques have provided strong evidence for active transport of sodium from lumen to blood with minimal movement of water, with a resulting positivity of the serosal surface of about 70 mV with respect to the lumen in the rat.[56] The potassium concentration within the lumen was influenced by the flow rate, being greater for slow flow rates. With fastest flow rates however its concentration leveled at 30 mM. Potassium transport was not electrogenic as in *Cecropia* midgut and cochlea. Its movement into the lumen might on the other hand have been brought about by electrogenic pumping of sodium into the blood, at least in the rat where the transepithelial potential was as high as 70 mV and favored passive potassium flux.

POTASSIUM TRANSPORT IN SWEAT GLANDS

These produce a hypo-osmotic solution in a manner analogous to the production of saliva, the primary secretion being modified during passage through a relatively water-impermeable duct. Sweat is also richer in potassium and poorer in sodium than interstitial fluid, due apparently to ouabain-sensitive transport of ions at the basal or serosal membrane.[57] The ratio of sodium uptake to potassium secretion seemed to be as great as 13.6:1, so the latter process may have been a passive response to electrogenic pumping of sodium ions from the tubular lumen to the blood. However in the rat's paw,[58] sweat glands seem to produce a secretion containing up to 150 mM K, and this may be a primary secretion since the ducts of the glands are very short.

RENAL POTASSIUM TRANSPORT

The handling of potassium ions by the kidney tubule has been more thoroughly investigated than any other form of epithelial transport and it is clearly of clinical significance in relation to the mode of action and the potassium-conserving properties of some diuretics.

In the proximal tubule the potassium that has been filtered at the glomerulus is reabsorbed by a process which may be active or passive,[59] depending on transepithelial potential and solvent drag. Since about 70% of the glomerular ultrafiltrate is reabsorbed in this segment, the influence of water movement is probably significant.

Most of the filtered potassium has been reabsorbed by the time the tubular fluid reaches the distal convoluted tubule, and the sodium concentration here may be 50–80 m.mole·l^{-1} and so is also hypo-osmotic. The potassium appearing in the urine must then be secreted in the distal nephron. The delicate regulation of potassium homeostasis is an important function of this segment; and from the lumen, sodium and water are reabsorbed in exchange for the potassium secreted.

The concentration of potassium in the luminal fluid at the end of the distal tubule depends on a number of factors, including the rate of flow within the lumen. For example, [K]$_L$ was 6 mM at a flow rate of 30 nl·min^{-1}, rising to 12 mM at 5 nl·min^{-1}, so the transport step was apparently not rate limiting under normal conditions.[60] The [K]$_L$ value at 5 nl·min^{-1} was the same as that observed under stop–flow conditions. The luminal potassium concentration [K]$_L$ was increased during K loading of the animal or during alkalosis induced by NaHCO$_3$ administration, both of which promoted potassium secretion in the distal tubule. Potassium deprivation and acidosis induced for example by hypercapnia inhibited potassium secretion, resulting in a fall of $[K]_L$ below the plasma level. Reabsorption of potassium from the lumen into the tubular cells was promoted under the latter conditions, and this appeared to be an active process.

A study of the kinetics of ^{42}K fluxes across the epithelium of single distal tubules of the rat *in vivo* under control conditions and those mentioned above has provided strong evidence that potassium secretion was largely determined by changes in peritubular transport of sodium and potassium. The technique involved the perfusion of the peritubular capillary network with ^{42}K Ringer and of the lumen of the distal tubule with tracer-free solution. From the steady-state flux measurements and from the rate of washout of tracer into the lumen after cessation of the peritubular perfusion with tracer, unidirectional fluxes, rate coefficients and the size of intracellular potassium pools were determined under control conditions and during stimulation and depression of tubular potassium secretion. The results of these studies are summarized in Table 8.2. Increased peritubular potassium uptake during loading of rats with this cation increased the cellular potassium pool, while potassium depletion reduced both. Sodium depletion also depressed potassium secretion, demonstrating the importance of sodium transport to the process.

$(Na^+ + K^+)$-ATPase activity has been found associated with the basolateral membranes in the early distal convoluted tubule, while K-dependent phosphatase activity seemed to be restricted to the cytoplasmic side of this membrane, thereby confirming the location of the Na–K pumping at the peritubular surface of the tubular cells. The concentration of potassium in the tubular cells as determined by this pump therefore seems to be a major determinant in the secretory process and also in the transepithelial potential as determined possibly by the electrogenic pumping of sodium ions from lumen to blood.

The presence of a potassium pump at the luminal surface of the cell has been indicated by the net reabsorption of this cation against a steep electrochemical potential gradient in rats on low-potassium diet and increase in luminal potassium concentration after administration of cardiac glycoside.[64]

There has been some disagreement about the magnitude of the transepithelial PD of the distal tubule. The use of standard microcapillary electrodes (tip diameter <1 μm) has given values ranging from -19.3 mV (lumen negative) in the early convoluted tubule to -47.2 mV in that part nearest the collecting duct.[65] Measurements made with larger tips[66] (3–5 μm) have indicated values of $+4$ mV and -20 mV, respectively, the difference being attributed to an artifact introduced by the finer tips.

Table 8.2 Summary of Mean Values of Kinetic Parameters from Distal Tubule of the Rat[a]

	Control	Low K	High K	5% NaHCO$_3$	Low Na
$m_{C,L}$	3.04 ± 0.397	1.66 ± 0.289	10.69 ± 2.05	7.63 ± 0.766	6.66 ± 0.875
$m_{C,I}$	6.76 ± 0.764	2.41 ± 0.344	16.7 ± 3.09	13.4 ± 1.32	12.7 ± 2.35
$m_{I,L}$	9.15 ± 2.11	4.16 ± 0.903	18.3 ± 2.6	18.6 ± 2.84	16.6 ± 3.55
$k_{I,C}$	0.59 ± 0.084	0.42 ± 0.070	0.82 ± 0.121	0.57 ± 0.042	0.63 ± 0.094
$k_{I,L}$	0.66 ± 0.131	0.64 ± 0.118	0.63 ± 0.119	0.75 ± 0.059	0.85 ± 0.151
$k_{L,I}$	29.10 ± 6.78	32.3 ± 7.37	25.2 ± 4.46	29.7 ± 4.5	85.5 ± 11.36
S_I	12.6 ± 1.32	6.55 ± 0.963	22.1 ± 3.65	24.5 ± 2.75	21.6 ± 3.17

[a]Subscripts C, I, and L refer to plasma (capillary), intracellular fluid, and tubular lumen, respectively; $m_{C,L}$ therefore is unidirection flux of ^{42}K from plasma to lumen (m.mole·min^{-1}·10^{-8}·mm^{-1}); $k_{I,C}$ is rate coefficient of ^{42}K transfer from tubular cell to plasma, and S_I is the cellular potassium pool (m.mole·min^{-1}).[61]

Double-barreled potassium-selective microelectrodes have been used[67] to measure the effective concentration of potassium in the epithelial cells of rat kidney under various conditions, and membrane potentials at the luminal and peritubular surfaces of the cell were also measured by means of conventional microelectrodes. Mean $[K]_i$ was 46.5 ± 1.6 m.mole·l^{-1} cell fluid under normal conditions, increasing to 60.5 ± 2.1 m.mole·l^{-1} under chronic K loading and falling to 36.5 ± 2.6 m.mole·l^{-1} in K depletion. In metabolic alkalosis and acidosis, $[K]_i$ values were 51.5 ± 0.9 and 38.7 ± 1.4 m.mole·l^{-1}, respectively. At the peritubular boundary of the cell, E_m was -67 mV, which was significantly more negative than the calculated E_K of -60.7 mV at this membrane, indicating that potassium ions should move passively into the cells. Because the membrane potential was not uniform along the luminal membranes of the distal tubule, E_m was approximately equal to calculated E_K (-24 mV) in the early part of the segment but was less negative than E_K by 4.5 mV in the later part of the segment, so that potassium ions should move passively out of the cells into the lumen.

Luminal membrane was depolarized in the kaliuretic state of potassium loading or alkalosis, but was hyperpolarized in the kaliupenic state of potassium depletion and acidosis. The hyperpolarization tended to reduce the transepithelial potential. The magnitude of this potential was therefore influenced by the potassium status of the animal, being about -85 mV in rats on K-rich diet[68] compared with -30 mV approximately during K deprivation. The rate of potassium secretion into the distal tubule was always less than the rate of sodium reabsorption but the coupling ratio was quite variable. Although in the majority of cases examined the secretion of K seemed to be passive, recent experiments[69] have indicated that in some cases at least energy may be required. In these studies transepithelial PD and potassium concentrations at both surfaces of the tubular cells were measured simultaneously in two groups of rats previously fed a normal diet. Group I was infused with KCl or K_2SO_4 solutions, while group II was infused with $NaHCO_3$ solution. The PD was measured with a microelectrode having a large tip diameter (3 μm), and tubular fluid was collected with micropipettes of 8 μm tip diameter. The infusion produced a tenfold increase in potassium excretion in group I and a threefold increase in group II. The initial PD was -46 mV in the former and -54 mV in the latter. During the collection of tubular fluid, the values fell to -23 and -30 mV, respectively. The tubular fluid/plasma [K] ratio averaged 5.0 and 4.4, which exceeded values predicted from the simultaneously measured PD, indicating that potassium had to be moved against an electrochemical potential gradient.

The intracellular potassium concentration of renal tubular cells measured indirectly appeared to be about 100–150 m.mole·l⁻¹ cell fluid,[70] and this range of values was generally accepted until direct measurements[6] with the K-selective microelectrodes indicated considerably lower values comparable to those found in toad bladder by the same technique.[15-17] The significance of these distinct sets of values is not clear, but it should be noted that the K-electrode measures the effective potassium concentration rather than the total concentration; and in toad bladder, where a_K^i was only 39.3 to 43.0 m.mole·l⁻¹, only a small fraction of total potassium seemed to exchange with ⁴²K within 60 min.

DIURETICS AND RENAL POTASSIUM MOVEMENT

Finally the effect of diuretics on renal potassium will be considered because of its clinical importance. The use of some diuretics has led to serious depletion of body potassium stores, while some newer drugs are relatively free of this unacceptable side effect. This difference may be traced to their sites of action through close examination of the factors affecting potassium movement in the renal tubule. Amiloride differs from other diuretics in its marked potassium-conserving properties,[71,72] which are very desirable during prolonged use. This widely used drug when applied to frog skin or toad bladder[12,13] appeared to act by decreasing the permeability of the apical membrane to sodium, thereby decreasing its passive movement into the cells and subsequent active transport across the serosal membrane on the Na–K pump. Its effect on the distal tubule of the rat kidney in the control state[73] was to change the transepithelial potential of the early part of the segment from +8.0 mV to +10.5 mV and that of the later part from −18 mV to +2.5 mV (measured with 3 μm tip electrodes). The change with small tipped electrodes was from −45.6 mV to −25.8 mV, thus creating an electrochemical potential gradient less favorable for potassium secretion, as in toad bladder.[74] It is difficult to know whether the hyperpolarization of the luminal membrane due to blocking of sodium leak by amiloride or reduced activity of the Na–K pump was the principal contributor to reduced secretion of potassium. The tubular fluid/plasma [K] ratio remained below unity in the presence of this diuretic, and collection and analysis of free flow sample of tubular fluid from superficial tubules of rat indicated complete suppression of tubular potassium secretion with only a small modification of sodium reabsorption in this segment.

As might be expected, the cardiac glycoside ouabain also acted as a diuretic and in animals on a low-K diet it induced both natriuresis and kaliuresis. The latter effect was independent of increased sodium delivery to the distal tubule but was believed to be due to inhibition of potassium reabsorption from the lumen of the distal tubule.

A third diuretic with yet another action is furosemide, which inhibited sodium and chloride transport in the loop of Henle,[76] resulting in the delivery to the distal tubule of a much greater fraction of fluid than normal. This in turn stimulated much greater sodium reabsorption at this segment, with consequent elevation of K secretion.

CONCLUSIONS

It is not possible within this short monograph to deal with all cases of epithelial transport of potassium. The main omissions are its transport in the gastrointestinal tract[77] and at the blood-brain barrier.[78]

Although potassium plays a minor role compared to that of sodium in terms of magnitude of transport and energy requirement at epithelial membranes, its relationship to the acid–base balance in the body is rather unique and merits separate treatment.

REFERENCES

1. H. H. Ussing & K. Zerahn, *Acta Physiol. Scand.* **23**, 110 (1951).
2. H. Linderholm, *Acta Physiol. Scand.* **31**, 36 (1954).
3. V. Koefoed-Johnsen & H. H. Ussing, *Acta Physiol. Scand.* **43**, 298 (1958).
4. F. Rawlins, L. Mateu, F. Fragachan & G. Whittembury, *Pflügers Arch.* **316**, 64 (1970).
5. D. Erlij & J. Aceves, *Biophys. J.* **9**, A163 (1969).
6. B. A. Robinson & A. D. C. McKnight, *J. Membr. Biol.* **26**, 269 (1976).
7. A. L. Finn, *Nature* (Lond.) **250**, 495 (1974).
8. J. W. Mills, S. A. Ernst & O. R. Dibona, *J. Cell Biol.* **73**, 88 (1977).
9. F. C. Herrera, *Am. J. Physiol.* **210**, 980 (1966).
10. P. M. Cala, N. Cogswell & L. J. Mandel, *J. Gen. Physiol.* **71**, 347 (1978).
11. S. A. Lewis, N. K. Wills & D. C. Eaton, *J. Membr. Biol.* **41**, 117 (1978).
12. S. A. Lewis, D. C. Eaton, C. Clausen & J. M. Diamond, *J. Gen. Physiol.* **70**, 427 (1977).
13. A. D. C. McKnight, M. M. Civan & A. Leaf, *J. Membr. Biol.* **20**, 365 (1975).
14. B. A. Robinson & A. D. C. McKnight, *J. Membr. Biol.* **26**, 239 (1976).
15. G. Kimura, S. Urakabe, S. Yuasa, S. Miki, Y. Takamitsu, Y. Orita & H. Abe, *Am. J. Physiol.* **232**, F196 (1977).
16. G. Kimura & M. Fujimido, *Jpn. J. Physiol.* **27**, 291 (1977).

References

17. J. deLong & M. M. Civan, *J. Membr. Biol.* **42**, 19 (1978).
18. G. Whittembury & F. Proverbio, *Pflügers Arch.* **316**, 1 (1970).
19. M. Cereijido & P. F. Curran, *J. Gen. Physiol.* **48**, 543 (1965).
20. A. L. Finn, *Physiol. Rev.* **56**, 453 (1976).
21. A. L. Finn & H. Nellans, *J. Membr. Biol.* **8**, 189 (1972).
22. R. Whittam & R. E. Davies, *Biochem. J.* **56**, 445 (1954).
23. A. L. Finn, *J. Membr. Biol.* **12**, 301 (1973).
24. M. Burg, L. Stoner, J. Cardual & N. Green, *Am. J. Physiol.* **225**, 119 (1973).
25. W. R. Harvey & S. Nedergaard, *Proc. Natl. Acad. Sci. U.S.* **51**, 157 (1964).
26. J. A. Haskell, W. R. Harvey & R. M. Clark, *J. Exp. Biol.* **48**, 25 (1968).
27. W. R. Harvey, J. A. Haskell & S. Nedergaard, *J. Exp. Biol.* **48**, 1 (1968).
28. W. R. Harvey, J. A. Haskell & K. Zerahn, *J. Exp. Biol.* **46**, 235 (1967).
29. J. A. Haskell & R. D. Clemons, *J. Cell. Biol.* **19**, 32 (1963).
30. S. Nedergaard & W. R. Harvey, *J. Exp. Biol.* **48**, 13 (1968).
31. L. Citron, D. Exley & L. S. Hallpike, *Br. Med. Bull.* **12**, 101 (1956).
32. C. A. Smith, O. H. Lowry & M. C. Wu, *Laryngoscope* **64**, 141 (1956).
33. C. G. Johnstone, R. S. Schmidt & B. M. Johnstone, *Comp. Biochem. Physiol.* **9**, 335 (1963).
34. P. M. Sellick & B. M. Johnstone, *Pflügers Arch.* **352**, 339 (1974).
35. P. M. Sellick & G. R. Bock, *Pflügers Arch.* **352**, 351 (1974).
36. I. Melchior & J. Syka, *Pflügers Arch.* **372**, 207 (1977).
37. I. Tasaki & C. S. Spyropoulos, *J. Neurophysiol.* **22**, 149 (1959).
38. W. Kuypers & S. E. Bonting, *Pflügers Arch.* **320**, 348 (1970).
39. W. Kuypers & S. E. Bonting, *Pflügers Arch.* **320**, 359 (1970).
40. S. Ruach, *J. Laryngol. Otol.* **80**, 1144 (1966).
41. Y. B. Choo & D. Tabowitz, *Ann. Otol. Rhinol. Laryngol.* **74**, 140 (1965).
42. B. M. Johnstone & P. M. Sellick, *Q. Rev. Biophys.* **5**, 1 (1972).
43. W. Kuypers & S. E. Bonting, *Biochim. Biophys. Acta* **173**, 477 (1969).
44. J. T. Y. Chou & K. Rodgers, *J. Laryngol. Otol.* **76**, 341 (1962).
45. J. L. Wood, P. S. Farrand & W. R. Harvey, *J. Exp. Biol.* **50**, 169 (1969).
46. L. H. Schneyer, J. A. Young & C. A. Schneyer, *Physiol. Rev.* **53**, 720 (1972).
47. O. H. Petersen & G. L. Petersen, *J. Membr. Biol.* **16**, 353 (1974).
48. A. Nishiyama & O. H. Petersen, *J. Physiol.* **242**, 173 (1974).
49. J. W. Putney Jr, B. A. Leslie & S. H. Marier, *Am. J. Physiol.* **235**, C128 (1978).
50. N. Iwatsuki & O. H. Petersen, *Pflügers Arch.* **377**, 185 (1978).
51. J. H. Poulsen & S. W. Bledsoe, *Am. J. Physiol.* **234**, E79 (1978).
52. L. P. Laugesen, J. O. D. Nielsen & J. H. Poulsen, *Pflügers Arch.* **364**, 167 (1976).

53. L. H. Schneyer, *Am. J. Physiol.* **217**, 1324 (1969).
54. J. H. Poulsen, M. Bundgaard & M. Møller, *Physiologist* **18**, 356 (1975).
55. Y. Imai, *J. Physiol. Soc. Jpn.* **27**, 304 (1965).
56. J. A. Young, E. Frömter, E. Schagel & K. E. Hamann, *Pflügers Arch.* **295**, 157 (1967).
57. J. Mangos, *Am. J. Physiol.* **224**, 1235 (1973).
58. S. W. Brushlow, K. Ikai & E. Gordes, *Proc. Soc. Exp. Biol.* **129**, 731 (1968).
59. F. S. Wright, *Kidney Int.* **11**, 415 (1977).
60. T. Morgan & R. W. Berliner, *Nephron* **6**, 388 (1969).
61. M. DeMello, G. Giebisch & G. Malnic, *J. Physiol.* **232**, 47 (1973).
62. J. Kyte, *J. Cell. Biol.* **68**, 287 (1976).
63. S. A. Ernst, *J. Cell. Biol.* **66**, 586 (1975).
64. M. Wiederholt, W. J. Sullivan & G. Giebisch, *J. Gen. Physiol.* **57**, 495 (1971).
65. G. Malnic & G. Giebisch, *Am. J. Physiol.* **223**, 797 (1972).
66. L. I. Barratt, F. C. Rector Jr., J. P. Kokko, C. C. Fisher & D. W. Seldin, *Kidney Int.* **8**, 368 (1976).
67. R. N. Khuri, S. K. Agulian & A. Kalloghlian, *Pflügers Arch.* **335**, 297 (1972).
68. F. S. Wright, Potassium Transport by the Renal Tubule, in *Kidney and Urinary Tract Physiology*, K. Thürau, Ed., University Park Press, Baltimore, 1974, pp. 79–106.
69. F. S. Wright, *Fed. Proc.* **35**, 465 (1976).
70. M. B. Burg & J. Orloff, *Am. J. Physiol.* **211**, 1005 (1966).
71. J. E. Baer, C. B. Jones, S. A. Spitzer & H. F. Russo, *J. Pharmacol. Therap.* **157**, 472 (1967).
72. E. A. Gambos, E. S. Freis & A. Maghandam, *N. Engl. J. Med.* **275**, 1215 (1966).
73. L. J. Barrett, *Pflügers Arch.* **361**, 251 (1976).
74. P. J. Bentley, *J. Physiol.* **195**, 317 (1969).
75. C. F. Duarte, F. Chomety & G. Giebisch, *Am. J. Physiol.* **221**, 632 (1971).
76. A. D. C. MacKnight, *Biochim. Biophys. Acta* **150**, 263 (1968).
77. A. D. C. MacKnight, *Kidney Intl.* **11**, 391 (1977).
78. C. F. Barnaby & C. J. Edmonds, *J. Physiol.* **205**, 647 (1969).
79. M. W. B. Bradbury & B. Stulkova, *J. Physiol.* **208**, 415 (1970).
80. E. M. Wright, *J. Physiol.* **226**, 545 (1972).
81. C. E. Johnson, D. J. Reed & D. W. Woodbury, *J. Physiol.* **241**, 359 (1974).

9
Cellular Potassium and Acid–Base Balance

The modification by acidosis and alkalosis of potassium secretion in the distal tubule, which mimicks the effects of potassium depletion and potassium loading, respectively, is mentioned in Chapter 8. During respiratory acidosis (hypercapnia) and metabolic acidosis (administration of NH_4Cl) the body responds by compensation through renal or respiratory mechanisms to bring the pH of the blood back to its normal value of 7.4. Tissues such as muscle, which occupies up to 40% of total body mass, also contributes significantly to the buffering of the acid load. In alkalosis similar mechanisms operate.

When respiratory acidosis is induced, the kidney tends to secrete hydrogen ions rather than potassium ions, and this may give rise to hyperkalemia. In dietary potassium depletion on the other hand the kidney conserves this cation, excreting hydrogen ions in preference, and this results in alkalosis.[1] Exchanges of hydrogen and potassium ions between tissues such as muscle and blood or interstitial fluid may also be responsible for such changes in the blood. Nephrectomized animals have been used[2-4] in order to examine such exchanges without the involvement of kidney. The alkalosis found in K depletion might have been due to migration of hydrogen ions into muscle fibers in exchange for potassium ions.[5-7] Acute administration of K^+ was found to result in a rapid fall in blood pH and HCO_3^-.[8]

When intracellular pH electrodes were introduced[9] into muscle fibers of mouse soleus, intracellular pH, pH_i, fell slowly by about 0.11 ± 0.03 unit when potassium was removed from the bathing medium, returning fairly rapidly to its normal value when K_0 was restored.

INTRACELLULAR pH OF MUSCLES AT REST

Before considering further the relationship between pH_0 and the movement of potassium between the muscle fiber water and interstitial fluid or blood, it might be profitable to examine the question of the distribution of hydrogen ions between the inside and outside of the cell and how it is controlled. On the assumption that the cell membrane was freely permeable to hydrogen ions, it has been suggested[10] that they are distributed passively across the cell membrane in a Donnan equilibrium, on which basis it was calculated that the pH_i should be about 6.0 at the normal pH_0 value of 7.4. Another prediction of this hypothesis was that the pH_i should change in response to addition or removal of external potassium, provided that the external fluid was well buffered.

One of the earliest methods used to determine pH_i was to measure or attempt to measure the distribution of bicarbonate ions between the muscle fiber water and blood. The HCO_3^- was determined[11] from the total acid-labile CO_2 of an alkaline extract of muscle. A correction was applied for a fraction of this which was not precipitated by $BaCl_2$ and which was believed not to be derived from HCO_3^-. These measurements yielded a pH_i value of 6.0, indicating passive distribution of hydrogen ions across the fiber membrane. Subsequent measurement of pH_i,[12-15] either directly with electrodes or from the distribution of weak acids or bases across the cell membrane have generally failed to confirm this low pH_i, a value of about pH 7 being more commonly found. A notable exception has been the finding[16] of a pH_i of 5.99 in normal resting leg muscles of rat *in situ*, by means of double-barreled pH-sensitive glass microelectrodes, insulated to within 5–20 μm of their tips, one barrel serving as reference electrode. A linear relationship was also found between pH_i and E_m when the latter was varied over the range of −30 to −90 mV. Using similar electrodes, it was later found[17] in muscles of rat and crab that measurements on superficial fibers gave lower pH_i values than measurements on deeper fibers which had values about pH 7. It was therefore concluded that the lower values were due to an artifact produced by inadequate insulation of the electrodes. If an inadequately insulated microelectrode was sampling pH_0 in addition to pH_i, it is not clear how measurements on superficial fibers gave pH_i values which were too low.

Criticism regarding insulation cannot be made however about the recessed tip type of microelectrodes[9] in which the pH-sensitive tip is surrounded by a microcapillary of borosilicate glass into which fluid must enter by capillary action (Fig. 2.3). With such an electrode pH_i was found to be 7.07 ± 0.007 at 37°C in fibers of mouse soleus bathed in physiological saline at 7.4 and bubbled with 5% CO_2. Changing pH_0 either by altering $[HCO_3^-]_0$ or the CO_2 level caused pH_i to change by 38.7% of the pH_0 change, the pH_i change being accomplished about 10 times faster on altering pCO_2.

A pH_i about 7.0 under normal conditions *in vivo* and *in vitro* would indicate that hydrogen ions were not passively distributed but had to be actively pumped out of the cell. Before considering possible mechanisms for such transport, we might consider what we mean by pH_i and also the reliability of the methods used to measure it. When a microelectrode tip, properly insulated, is introduced into a relatively large muscle fiber such as that of crab, with low to moderate metabolic rate, it may very well give a true indication of the hydrogen ion concentration in the sarcoplasm. It need hardly be stressed however that the cell interior may be quite heterogeneous and compartmentalized. It is possible for example that mitochondria which are known to actively pump out protons may create local high concentrations of these ions, and tissues such as mammalian diaphragm and soleus may have layers of these particles (see Plate 10) just below the sarcolemma.

We might therefore question the existence of a single homogeneous equilibrium state for this cation, particularly in smaller cells. The indirect measurement of pH_i from the distribution of weak acids and bases, such as DMO, across the cell membrane must also be subject to some uncertainty arising from the need to obtain a precise value for water distribution within the tissue. However in spite of this difficulty, good agreement seems to have been found[18] between pH_i measurements in single crustacean muscles fibers by the DMO method on the one hand and by direct intracellular electrode measurements on the other in both acid and alkaline media. Results obtained in these experiments also led to the conclusion that hydrogen ion distribution across the cell membrane varied with membrane potential but not in accordance with a simple Donnan equilibrium. There appeared to be two independent net fluxes of protons across the membrane, one representing passive diffusion and the other, the active transport on the pump. Although one might have expected hydrogen ions to be passively distributed in the resting state if their

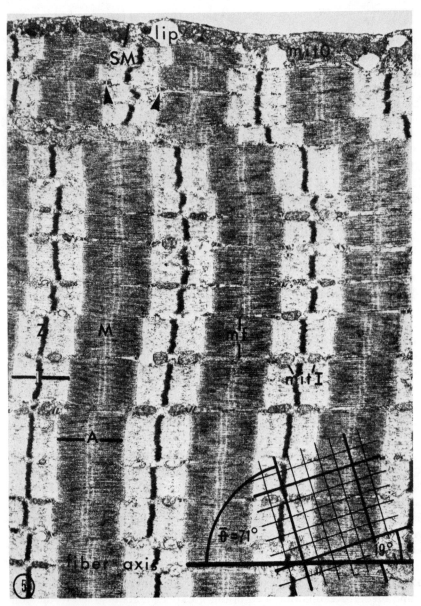

Plate 10 Longitudinal section through part of a soleus muscle fiber of the guinea pig. The lipid droplets (lip) and mitochondria (mito) lie close to the sarcolemma (SM). Arrowheads point to triads located at the junction of the A and I bands, between some but not all myofibrils (mf). The mitochondria in the I band (mitI) are often paired and A-band mitochondria are sparse. A portion of the test grid is shown oriented at the optimal angle $\theta = 19°$ and $71°$ to the fiber axis. Light-line spacing $d = 0.4\ \mu m$, ×12,000. From B. R. Eisenberg, A. M. Kuda & J. B. Peter, *J. Cell Biol.* **60,** 732 (1974).

permeability is much greater than that of potassium,[19] these cations are unique in their large degree of association with the nondiffusible anionic groups within the cell which accounts for the high buffering capacity found there and may increase the time needed for achievement of equilibrium after a change in external pH. Where equilibrium must be reached rapidly, as in the red blood cell, we find that a Donnan equilibrium exists with respect to these ions,[20] and a measured membrane potential[21] of -14 mV is in line with the following concentration ratios:

$$\frac{[H^+]_i}{[H^+]_0} = \frac{[Cl^-]_0}{[Cl^-]_i} = \frac{[HCO_3^-]_0}{[HCO_3^-]_i} \qquad (50)$$

In the human erythrocyte the potassium equilibrium potential would be about -80 mV, so these cations are evidently not passively distributed but must be accumulated actively by the cell.

Intracellular pH is believed to be maintained within fairly narrow limits under normal conditions by two processes, namely, (a) the active transport of HCO_3^- into the cell in exchange for chloride ions,[22,23] and (b) transport of H^+ out of the cell in exchange for potassium[24] or sodium ions.[25] It has also been suggested[26] that hydrogen ions might be transported out of the cell on the outward limb of the Na–K pump in muscle. In the study of active transport in relation to maintenance of hydrogen ion homeostasis of the cell, methods have been worked out for bringing about rapid changes in pH_i, that is, acid or alkaline loading of the cell. These changes have been conveniently studied with recessed-tip pH-sensitive microelectrodes. Both CO_2 and NH_3 pass readily through the lipid phase of the membrane and facilitate the achievement of equilibrium in response to a sudden change in pH_0.[27] An immediate and large fall in pH_i was found,[28] for example, in squid axons when the external medium was bubbled with 100% CO_2. Transient changes in pH_i were also seen in snail neurons[29] and in squid axons[30] on exposure of CO_2 and NH_3 where pH_0 was changed. For example, when NH_4Cl was added to the fluid bathing squid axons at pH 7.7, the pH_i increased rapidly as NH_3 entered the axoplasm and combined with H^+ to form NH_4^+. Some NH_4^+ also diffused slowly into the axon down its electrochemical potential gradient. On replacing the external fluid with fresh seawater, pH_i then fell rapidly from about 7.7 to 7.0, as NH_4^+ within the axon dissociated and NH_3 diffused out. The longer the period of exposure to the alkaline NH_4Cl, the lower the pH_i reached on returning to seawater.

In mouse soleus muscle fibers bathed in physiological saline at pH 7.4 and 37°C and bubbled with 5% CO_2, the pH_i was 7.07. When pCO_2 outside the fibers was reduced to zero, keeping pH_0 constant by decrease in $[HCO_3^-]_0$, pH_i increased within 1 min from 7.07 to about 7.3, and this was

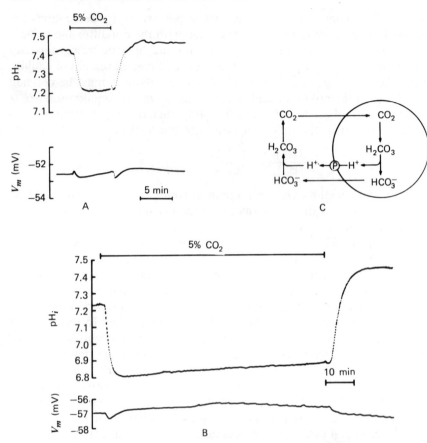

Figure 9.1 A, Effect of short-term exposure to CO_2 on pH_i and E_m. During the time indicated the axon was exposed to 5% CO_2 at constant pH_o (7.70). Note slight overshoot of pH_i on return to normal seawater. B, Effect of long-term exposure to same conditions. Note slow alkalinization during plateau phase and exaggerated overshoot on return to seawater. C, Inward movement of protons by the CO_2/HCO_3^- couple during the plateau phase. From W. Boron & P. DeWeer, *J. Gen. Physiol.* **67**, 91 (1976).

followed by a slow acidification over a period of 15–30 min at which time pH_i stabilized at 7.23 ± 0.03. Increase in pCO_2 at constant pH_o caused a rapid fall in pH_i with slow and partial recovery. The slowness of the latter response is illustrated in Fig. 9.1, where squid axons were exposed[27] for short (5 min) and long (1 hr) periods to 5% CO_2 while immersed in alkaline seawater at pH 7.7. After an initial rapid fall in pH_i the recovery to higher pH_i was still evident 1 hr later. On removing the CO_2 an overshoot occurred, with the pH_i rising above its resting value of 7.23 to over 7.4.

The initial acidification here was probably due to CO_2 entering the axon, being hydrated to carbonic acid which then dissociated. The final pH reached would depend to a large extent on the buffering power of the cell. The high lipid solubility of CO_2 makes it likely that it would reach the same concentration inside and outside the axon, so at equilibrium the following relationship should hold:

$$E_{H^+} = \frac{RT}{F} \ln \frac{[H^+]_o}{[H^+]_i} = E_{HCO_3} = \frac{RT}{F} \ln \frac{[HCO_3^-]_i}{[HCO_3^-]_o} \quad (51)$$

If however $E_{HCO_3^-}$ is less negative than E_m, this anion will diffuse out of the cell, thereby providing a sink for further entry of CO_2. One would be inclined to predict therefore a drift toward lower pH_i rather than the increase which was found and which may be attributed to active transport of H^+ out of the axon.

The calculated buffering power of the fibers of mouse soleus muscle[31] was found to range from 45.0 to 58.0 m.mole·pH unit^{-1}·l^{-1}, compared with about 76.6 ± 13.6 m.mole·l^{-1} in rat ventricle.[32]

ACTIVE TRANSPORT AND CONTROL OF pH_i

When giant nerve axons of squid were treated with cyanide or DNP, no recovery of pH_i occurred after loading with acid, whereas recovery took place both before addition of inhibitors and after their removal. Dialysis of the axons was carried out[22] to examine the recovery process. When ATP-free fluid containing cyanide was used in the dialysis tube, ATP_i was reduced to a very low level and pH_i remained low after an acid load until ATP was reintroduced.

Another surprising observation was that when the inhibitor of passive chloride fluxes in erythrocytes, 4-acetamido-4'-isothiocyanostilbene-2,2'-disulfonic acid (SITS),[33] was added to the external fluid, no recovery of pH_i occurred, even when ATP was present. The suggested mechanism of recovery here[23] was the active uptake of HCO_3^- coupled to Cl^- efflux. When $[Cl]_i$ was greatly reduced by internal dialysis with Cl-free fluid, recovery of pH_i after acid loading was also blocked. Finally, by introducing ^{36}Cl into the axon, a 50% increase in Cl efflux was demonstrated when pH_i was reduced to stimulate the HCO_3–Cl pump. In snail neurons also the pH_i-regulating system seemed to be inhibited when $(HCO_3^-)_o$ was removed.

The fact that addition of SITS had only a slight effect on recovery of mouse soleus muscle from an acid load[9] suggested that other mechanisms were available for regulating pH_i. It had been found[34] that ouabain produced a slight fall in pH_i of anterior gracilis muscles when measured *in vivo* in nephrectomized rats, while at the same time bringing the H^+ gradient more into line with that predictable from a Donnan equilibrium. It was suggested that the Na–K pump might also be able to transport H^+ out of the cell, which was in keeping with the observation[35] that sodium efflux was reduced by a fall in pH_i. However in the case of mouse soleus muscles[9] removal of K_0 did not seem to inhibit recovery from an acid load, while on the other hand the reduction of $[Na]_0$ by 11% greatly reduced recovery. On the basis of this finding it was concluded that H^+ ions were extruded from muscle fibers on a Na–H exchange pump making use of the electrochemical potential gradient of sodium ions across the membrane as an energy source. The available energy would therefore be reduced by reduction of $[Na]_0$ or by increase in $[Na]_i$, and the latter might be expected to occur in the presence of ouabain, which might explain the ouabain sensitivity of H^+ pump. The Na–K pump might also be activated by an increase in $[Na]_i$ arising from an exchange of Na^+ and H^+ across the cell membrane in response to an acid load. One might ask what effect such activation would have on intracellular potassium concentration.

CELLULAR POTASSIUM IN ACIDOSIS AND ALKALOSIS

Increasing pCO_2 appeared to hyperpolarize mammalian skeletal muscle fibers. For example, in mouse soleus,[31] E_m was -57.3 ± 1.5 mV in CO_2-free Ringer, -67.7 ± 1.3 mV in 1.63% CO_2, -71.8 mV in 5% CO_2, and about -73 mV in 7.1% CO_2. In rat EDL muscles[36] *in vitro*, E_m increased from -77.2 ± 0.2 mV in 5% CO_2 (pH_0 7.4) to -80.3 ± 0.2 mV in 8% CO_2 (pH_0 7.12), while at the same time potassium ion activity measured by cation-selective microelectrodes increased, with $[K]_i$ rising from 146.8 ± 3.9 m.mole·l^{-1} fiber water to 182.7 ± 7.3 m.mole·l^{-1} fiber water over a period of 1–2 hr. Under the same conditions E_m of soleus muscles decreased from -76.9 ± 0.1 mV to -72.2 ± 0.2 mV with $[K]_i$ falling from 139.7 ± 4.5 mV to 134.5 ± 3.8 mV, an insignificant change. When Krebs fluid bathing isolated rat muscles was bubbled with 14% CO_2 (pH_0 6.8), E_m fell to about -66 mV in both EDL and soleus muscles, while $[K]_i$ decreased to 107.5 ± 4.8 and to 111.5 ± 5.0 m.mole·l^{-1} fiber water, respectively, with a corresponding rise in $[Na]_i$. It seems possible that there may be a direct relationship between membrane potential and intracellular potassium concentration.

The finding of potassium uptake in EDL muscles in moderate acidosis and loss in more severe acidosis might indicate that the former was associated with the processes by which the fibers extruded hydrogen ions, while the latter might indicate the activation of this mechanism by excessive acid load. If for example hydrogen ions were extruded in exchange for sodium ions entering the fibers, then the active transport of the latter out of the fibers once more through the Na–K pump might be responsible for the potassium uptake. There would also seem to be a critical value of pH_0 below which this mechanism might fail with loss of regulation of intracellular pH. When pH_i of rat diaphragm was plotted against pH_0 over a wide range, induced by altering pCO_2 and HCO_3^-, a plateau of pH_i was evident,[15,37,38] extending from about pH_0 7.0 to 7.4. Below this range and also in the more alkaline region, the two pH values were linearly related. The loss of potassium below pH 7.0 was therefore understandable in terms of the disappearance of anionic charges on nondiffusible cellular constituents.

Net uptake of potassium by cardiac muscle, namely, by mammalian ventricle, has been reported during respiratory and metabolic acidosis.[39-42] In one case[42] the effect of hypercapnia on potassium movement was attributed to the action of catecholamines, since the administration of propranolol, a β-adrenergic blocker, changed the response to increased pCO_2 from net uptake to net loss of potassium. Stimulation of the sodium pump by catecholamines was much more pronounced in tonic muscles such as soleus than in phasic muscles such as the EDL, so it is perhaps surprising that it was the former which lost potassium in moderate acidosis. It should also be mentioned however that exposure to high pCO_2 increased the efflux of ^{22}Na from barnacle muscle fibers[43] and that this effect was reversed by the application of ethacrynic acid to the inside of the membrane.

The direction of potassium movement in experiments which were carried out *in vivo*[40,42] was determined from the A–V difference in potassium concentration following exposure to high CO_2 and after return to normal condition, at which time potassium appeared to leave the cardiac muscle. The results were also confirmed[39] by direct measurements of tissue potassium *in vivo* under respiratory and metabolic acidosis. Simultaneous measurements in rat diaphragm[39] indicated loss of potassium during acidosis. In keeping with potassium movements in myocardium, pH_i seemed to fall less than in skeletal muscles for a unit change in pH_0,[44] and measurements with recessed-tip microelectrodes also indicated better buffering in cardiac than in skeletal muscles.[32] This has been explained in terms of a more active proton pump in the former muscles.[45] If this pump is dependent on the electrochemical potential gradient for sodium ions across the cell membrane, then ultimately it may be the effectiveness of the Na–K pump in the tissue that determines whether potassium is gained or lost in respiratory and metabolic acidosis.

The contrasting behavior of EDL and soleus muscle of rat with respect to potassium movement during moderately increased pCO_2[36] may be due to a higher electrochemical potential gradient for sodium ions in the former, where $[Na]_i$ of the freshly dissected muscles were 14.0 ± 0.4 and 22.6 ± 0.5 m.mole·l^{-1} fiber water, respectively,[46] and membrane potentials measured *in vivo* under ideal conditions were -92.0 ± 0.6 and -84.5 ± 1.2 mV, respectively.[47,48]

It has been suggested[49] that potassium ions might move from the extracellular to the intracellular fluid *in vivo* when CO_2 tension was increased and in the reverse direction when tension was lowered, on the assumption that tissue was less well buffered than blood, so that the ratio $[H]_i/[H]_0$ increased in hypercapnia, potassium entering the cells to increase the ratio $[K]_i/[K]_0$ thereby restoring the Donnan equilibrium. Movement of potassium ions out of muscle fibers *in vitro* on exposure to high CO_2 has on the other hand been attributed to physiological saline having a lesser buffering power than the tissue, resulting in a fall in the ratio $[H]_i/[H]_0$. Ventilation of anesthetized dogs with 30% CO_2, which brought pH_0 below 6.8, caused a loss of potassium from gastrocnemius muscles *in vivo*[40] under conditions where cardiac muscle accumulated potassium and released it again in the posthypercapnic period. A rise in potassium measured on the basis of fat-free dry weight was seen in rabbit cardiac muscle in both metabolic and respiratory acidosis, which was inversely related to the pH gradient ($pH_0 - pH_i$) but showed no significant correlation with pH_i. In neither metabolic nor respiratory alkalosis was there a significant change in total potassium of the tissue.

The interpretation of changes in cellular potassium concentration in response to pH_0 or pH_i is made more difficult *in vivo* by processes of compensation which may result in changes in the concentrations of potassium, sodium, chloride, or bicarbonate in the blood plasma. For this reason nephrectomy was sometimes carried out. But this may also produce changes, including a rise in plasma potassium and fall in sodium,[50] with similar changes in liver and muscle. Metabolic acidosis with some respiratory compensation may be present. Significant perhaps was the finding that the membrane potential of anterior gracilis muscles decreased from -91 mV to -77 mV in spite of the increase in $[K]_i$. In liver however E_m changed from -43 mV to -48 mV, and this was believed to be due to activation of the electrogenic sodium pump, perhaps as a result of increased plasma potassium concentration. It would seem likely therefore that *in vitro* experiments where $[K]_0$ and the external concentration of other ions may be kept constant during acidosis may facilitate interpretation of results.

Finally if, as in neural tissue, the active exchange of HCO_3^- for Cl^-

across the muscle membrane contributed to the recovery process to a significant extent during hypercapnia, then the resulting fall in Cl_i by increasing the ratio $[Cl]_o/[Cl]_i$ might account for the hyperpolarization and uptake of potassium seen in EDL and cardiac muscle fibers.

REFERENCES

1. R. B. Tobin, *Am. J. Physiol.* **186**, 131 (1956).
2. R. O. H. Irvine & J. Dow, *Clin. Sci.* **31**, 317 (1966).
3. R. O. H. Irvine & J. Dow, *Metabolism* **15**, 1011 (1966).
4. R. E. Cooke, W. E. Segar, D. B. Cheek, F. E. Coville & D. C. Darrow, *J. Clin. Invest.* **31**, 798 (1952).
5. D. A. K. Black & M. D. Milne, *Clin. Sci.* **11**, 397 (1952).
6. L. I. Gardner, E. A. MacLachlan & H. Berman, *J. Gen. Physiol.* **36**, 153 (1952).
7. S. J. Saunders, R. O. H. Irvine, M. A. Crawford & M. D. Milne, *Lancet* **468** (1960).
8. J. B. Hudson & A. S. Relman, *Am. J. Physiol.* **203**, 209 (1962).
9. C. C. Aickin & R. C. Thomas, *J. Physiol.* **273**, 295 (1977).
10. P. J. Boyle & E. J. Conway, *J. Physiol.* **100**, 1 (1941).
11. E. J. Conway & P. J. Fearon, *J. Physiol.* **103**, 274 (1944).
12. P. C. Caldwell, *J. Physiol.* **142**, 22 (1958).
13. P. G. Kostyuk & Z. A. Sorokina, "On the mechanism of hydrogen ion distribution between cell protoplasm and the medium" in *Membrane Transport and Metabolism*, A. Kleinzeller & A. Kotyk, Eds. Academic Press, New York, 1960, pp. 193-203.
14. W. J. Waddell & T. C. Butler, *J. Clin. Invest.* **38**, 720 (1959).
15. S. Adler, A. Roy & A. S. Relman, *J. Clin. Invest.* **44**, 8 (1965).
16. N. W. Carter, F. C. Rector Jr., D. S. Campion & D. W. Seldin, *J. Clin. Invest.* **46**, 920 (1967).
17. M. Paillard, *J. Physiol.* **223**, 297 (1972).
18. J. A. M. Hinke & M. R. Menard, *J. Physiol.* **262**, 533 (1976).
19. J. W. Woodbury, Fluxes of H^+ and HCO_3^- Across Frog Skeletal Muscle Cell Membranes, in *Ion Homeostasis of the Brain*, B. K. Siesjö & S. C. Sørensen, Eds., Alfred Benzon Symposium III, Munksgaard, Copenhagen, 1971, pp. 270-289.
20. J. Funder & J. O. Wieth, *Acta Physiol. Scand.* **68**, 234 (1966).
21. U. V. Lassen & O. Sten-Knudsen, *J. Physiol.* **195**, 681 (1968).
22. J. M. Russell & W. F. Boron, *Nature* (Lond.) **264**, 73 (1976).
23. R. C. Thomas, *Nature* (Lond.) **262**, 54 (1976).

24. R. O. H. Irvine, S. J. Saunders, M. D. Milne & M. A. Crawford, *Clin. Sci.* **20**, 1 (1960).
25. R. C. Thomas, *J. Physiol.* **273**, 317 (1977).
26. J. W. Woodbury, Regulation of pH, in *Physiology and Biophysics*, T. C. Ruch & H. D. Patton, Eds., Saunders, Philadelphia, 1965, pp. 899–934.
27. P. DeWeer, *Resp. Physiol.* **33**, 41 (1978).
28. P. C. Caldwell, *J. Physiol.* **142**, 22 (1958).
29. R. C. Thomas, *J. Physiol.* **238**, 159 (1974).
30. W. F. Boron & P. DeWeer, *J. Gen. Physiol.* **67**, 91 (1976).
31. C. C. Aickin & R. C. Thomas, *J. Physiol.* **267**, 791 (1977).
32. D. Ellis & R. C. Thomas, *J. Physiol.* **262**, 755 (1976).
33. P. A. Knauf & A. J. Rothstein, *J. Gen. Physiol.* **58**, 190 (1971).
34. J. A. Williams, C. D. Withrow & D. M. Woodbury, *J. Physiol.* **212**, 101 (1971).
35. R. D. Keynes, *J. Physiol.* **178**, 305 (1965).
36. P. Elsner, R. P. Kernan & M. MacDermott, *J. Physiol.* **285**, 47P (1978).
37. N. Heisler, *Resp. Physiol.* **23**, 243 (1974).
38. W. J. Waddell & R. G. Bates, *Physiol. Rev.* **49**, 285 (1968).
39. P. A. Poole-Wilson & I. R. Cameron, *Am. J. Physiol.* **229**, 1305 (1975).
40. R. I. Lade & E. B. Brown, Jr., *Am. J. Physiol.* **204**, 761 (1963).
41. J. C. Mithoefer, H. Kazemi, F. D. Holford & I. Friedman, *Resp. Physiol.* **5**, 91 (1968).
42. D. G. Spiker & C. W. Smith, *Circ. Res.* **30**, 535 (1972).
43. B. G. Danielson, E. E. Bittar, S. Chen & E. Tong, *Life Sci.* **10**, 1225 (1971).
44. R. L. Clancy & E. B. Brown, *Am. J. Physiol.* **211**, 1309 (1966).
45. Y. L. Lai, B. A. Attebery & E. B. Brown, *Resp. Physiol.* **19**, 123 (1973).
46. F. A. Sreter & G. Woo, *Am. J. Physiol.* **205**, 1290 (1963).
47. R. P. Kernan & M. MacDermott, *J. Physiol.* **233**, 363 (1973).
48. R. P. Kernan & I. McCarthy, *J. Physiol.* **226**, 62P (1972).
49. W. O. Fenn & T. Asano, *Am. J. Physiol.* **185**, 567 (1956).
50. J. A. Williams, C. D. Withrow & D. M. Woodbury, *J. Physiol.* **212**, 117 (1971).

INDEX

4-Acetamido-4'-isothiocyanostilbene-2,2'
  disulfonic acid:
    effect on chloride fluxes, 181
    effect on erythrocytes, 181
Acetylcholine:
    effect on potassium conductance in
      muscle, 71
    effect on salivary glands, 166
Acidosis:
    metabolic, 175
      in cardiac muscle, 183
    in muscle:
      potassium loss, 182-183
      potassium uptake, 182-183
    respiratory, 176
      in cardiac muscle, 183
      response of kidney, 175
Acinar cells, see Salivary glands
Action potential, 58-83
    in cardiac muscle, 70-80
      deviation from Hodgkin-Huxley
        equations, 73
      role of potassium, 70-80
      role of sodium, 70-80
    in nerves:
      gating mechanism, 59-61
      Hodgkin-Huxley model of conductance
        control, 59-61
      role of potassium, 58-83
      role of sodium, 58-83
      squid axon, 58-59
      theoretical, 61
    in Purkinje fibers, 75-80
    in skeletal muscle, 62-66
      gating mechanism, 59-61
      role of calcium, 62-66
      role of potassium, 62-66
      role of sodium, 62-66
Active transport:
    in algae, 5-10
    application to Michaelis-Menten kinetics,
      12-14
    in bacteria, 114-115
    in cardiac tissue, 106-107
      effect of cardiac glycosides,
        106-107
    in cochlea, 160-165
    in dialyzed squid axon, 109-112
    in erythrocytes, 85-86, 181
    in *Escherichia coli,* 141
    in frog skin, 151
    in *Griffithsia,* 9
    in *Halicystis ovalis,* 10
    in *Halicystis parvula,* 10
    in liver, 112
    in mitochondria, 120-135

Index

Active transport (Cont'd)
  in muscle:
    in acidosis, 182-185
    in alkalosis, 182-185
  in *Neurospora crassa*, 11
  in *Nitella translucens*, 6-7
  in nonepithelial cells, 85-115
  in rabbit bladder, 155
  in resting muscle, 179
  in root hairs, 12-14
  in salivary glands, 166
  in skeletal muscle, 100-102
    effect of innervation, 103-104
  in smooth muscle, 107-109
  structure of transport protein, 86-88
  in sweat glands, 167
  in toad bladder, 152-153
  in yeast, 11, 112-114, 141
  see also $(Na^+ + K^+)$-ATPase; Transepithelial transport
Activity coefficient:
  intracellular potassium, 37
  intracellular sodium, 37
Adenosine triphosphate:
  generation in mitochondria, 124-131
  role in sodium-potassium pump, 89-92
S-Adenosylmethionine synthetase:
  potassium requirement, 147
Adipose tissue:
  potassium concentration, 18, 24
Adrenaline, 55
  stimulation sodium-potassium pump, 105
alpha-Adrenergic receptors:
  in salivary glands, 165
Aerobacter aerogenes:
  purine synthesis:
    potassium requirement, 145
Afterpotential:
  in skeletal muscle, 63
Aldolase:
  in bacteria, 146
  in muscle, 146
  potassium requirement, 146
  in yeast, 146
Algae:
  freshwater:
    cytoplasmic potassium concentration, 5
    vacuolar potassium concentration, 5
    see also Chara australis; Chara corallina; Nitella translucens
  marine:
    cytoplasmic potassium concentration, 5
    vacuolar potassium concentration, 5
    see also Griffithsia; Halicystis ovalis; Halicystis parvula
  potassium uptake, 5-10
  in illuminated cells, 9
Alkalosis, 175
  in potassium depletion, 175-176
Amiloride:
  effect on epithelial tissue, 155
  potassium-conserving properties, 171
Amino acyl-*s*-RNA:
  in protein synthesis, 142
  potassium dependence, 142
Aminopyridine:
  blocking of potassium channels, 53
Ammonia:
  effect on intracellular pH, 179
Anesthetics, local:
  effect on skeletal muscle, 103
Anomalous rectification:
  in Purkinje fibers, 75-76
  in skeletal muscle, 41, 45-51
    hypotheses to explain, 46-47
    role of surface membrane, 50
    role of T-system, 47-50
Arsenite:
  effect on sodium-potassium pump, 91
Association-induction hypothesis for K^+ accumulation, 31-38
  effect on activity coefficients, 37
  experimental evidence for, 33-38
  potassium-binding proteins, 31-33
  role of ATP, 31
  role of cell water, 31
$(Na^+ + K^+)$-ATPase, 147
  binding of cardiac glycosides, 107
  catalytic protein, 87-88
  in cochlea, 163
  conformational change, 88
  direction of cation movement, 89
  in epithelial tissue, 155

$(Na^+ + K^+)$-ATPase (Cont'd)
   inhibition by ouabain, 88
   inhibition by strophanthidin, 89
   in kidney, 168
   labelling with iodonaphthyl-1-azide, 88
   phospholipid dependence, 88
   phosphorylation of membrane protein, 89
   proteolysis, 87-88
   in rabbit bladder, 155
   ratios of cation movement, 89
   in red blood cells, 89-100
   from renal medulla, 87-88
   role of ATP, 89-92
   in salivary glands, 166
   sialoglycoprotein, 87-88
   stimulation by insulin, 105
   stimulation by isoprenaline, 106
ATPases:
   potassium requiring, 144
Atrium muscle of heart:
   action potential, 73

Bacillus amyloliquifaciens:
   potassium dependence of protein synthesis, 140
Bacillus subtilis:
   potassium dependence of protein synthesis, 140
Bacteria:
   active transport, 114-115
   intracellular potassium concentration, 3
   phosphate uptake, 114
   potassium uptake, 3, 114-115
   regulation of osmotic pressure, 114
   see also *Escherichia coli*
Barium:
   blocking of potassium channels, 52-54
Bicarbonate:
   active transport, 179
   in erythrocytes, 181
   role in intracellular pH, 176-181
Bicarbonate-chloride pump, 181
Biopsy:
   skeletal muscle, 18-20
Biotite, 1
   weathering of, 1
Bladder, rabbit:
   active transport, 155
   $(Na^+ + K^+)$-ATPase, 155
   intracellular potassium activity, 154
   ion permeabilities, 154
Bladder, toad:
   active transport, 152-153
   compartmentalization of epithelial tissue, 159
   effect of ethacrynic acid, 156-157
   effect of K^+-free bathing fluid, 157
   effect of ouabain, 153, 157
   effect of rotenone, 157
   granular cells, 152-153
   mitochondria-rich cells, 152-153
   potassium fluxes, 159
   potassium permeability, 153
   voltage steps across, 153
Blood, whole:
   potassium concentration, 24
Bone:
   potassium concentration, 24
Brain:
   potassium concentration, 24

Caesium:
   blocking of potassium channels, 51-53
   effect on Purkinje fibers, 79-80
Calcium:
   and action potential:
     in cardiac muscle, 70-80
   effect on potassium conductance, 54-55
   effect on gating, 80-82
   effect on potassium distribution, 8
Carbachol:
   effect on salivary glands, 165
Cardiac glycosides:
   diuretic action, 171
   inhibition of sodium-potassium pump, 106-107
   inotropic effect, 106-107
   strophanthidin, 107
   see also Ouabain
Cardiac muscle:
   action potential, 70-80
     in atrium, 73
     compared to skeletal muscle, 70

Cardiac muscle (Cont'd)
    effect of acetylcholine, 71
        in pacemaker cells, 73
        plateau, 71
    active transport, 106-107
    application of Hodgkin-Huxley equation, 73
    compartmentalization, 70
    effect of cardiac glycosides, 106-107
    effect of propranolol, 183
    metabolic acidosis, 183
    potassium current, 70-80
    repolarization, 70-80
    respiratory acidosis, 183
    trabeculum, 70
    see also Purkinje fibers
Cecropia, midgut tissue:
    potassium pump, electrogenic, 114
        location, 164-165
    potassium transport, 160, 163-165
        effect of anoxia, 160, 163-165
        effect of ouabain, 160
        inhibition by DNP, 160
Cell membrane:
    integral proteins, 86
    see also $(Na^+ + K^+)$-ATPase
    K^+-activated phosphatase, 88
Cerebrospinal fluid:
    ion concentrations, 160
Chara australis:
    intracellular potassium:
        activity coefficient, 37-38
Chara corallina:
    partial ion conductances, 8-9
    passive accumulation of potassium, 8
    total membrane conductance, 8-9
Chemiosmotic hypothesis, 124-131
Chloride:
    active transport:
        in *Halicystis ovalis,* 10
        in *Nitella translucens,* 6-7
    bicarbonate-chloride pump, 181
    fluxes:
        in erythrocytes, 181
        inhibition by SITS, 181-182
Cholinergic receptors:
    in salivary glands, 165
Cochlea:
    active transport, 160-165
        $(Na^+ + K^+)$-ATPase, 163
        effect of anoxia, 163-164
        effect of ethacrynic acid, 162-163
    electrogenic potassium pump, 162
    electrogenic sodium pump, 162
    endocochlear potential, 161-162
    fluids:
        ion concentrations, 161
    potassium transport, 160-165
    sodium fluxes, 162-163
    sodium-potassium pump, 162
        location, 163-164
    structure, 161
Cyanide:
    effect on intracellular pH, 181

Dehydrogenases:
    potassium requiring, 144
Delayed rectification:
    in skeletal muscle, 62-63
Diabetes mellitus:
    decrease in intracellular potassium, 17, 20
Diactin, 148
2,3-Dimercaptopropanol:
    effect on sodium-potassium pump, 91
Dinitrophenol:
    action in mitochondria, 132-135
    effect on *Cecropia* midgut tissue, 160
    effect on intracellular pH, 181
Diuretics, 167
    amiloride, 171
    blockage of sodium transport:
        in kidney, 156
    effect on potassium fluxes in kidney, 171-172
    furosemide, 172
    ouabain, 171
    potassium-conserving, 171
    potassium-depleting, 171
Donnan equilibrium:
    of permeable ions across cell membrane, 38

Electrogenic pump, *see* Potassium pump, electrogenic; Sodium pump, electrogenic

Endolymph:
  ion concentrations, 160
Enzymes:
  changes during potassium deprivation, 143-145
  potassium requiring, 138-139, 145, 146-148
    mammalian, 143-145
    membrane-bound, 147
Epithelial tissue:
  active transport, 151-172
  apical surface, 151
    effect of nystatin, 154-155
    potassium fluxes, 159
  compartmentalization, 159
  effect of vasopressin, 158
  electrogenic sodium pump, 154-159
  intracellular potassium activity, 154
  ion permeabilities, 154
  neutral pump, 154-159
  potassium fluxes, 159
  potassium-potassium exchange, 159
  serosal surface, 151
    $(Na^+ + K^+)$-ATPase, 155
    effect of amiloride, 155
    effect of nystatin, 154-155
    potassium fluxes, 159
  see also Transepithelial transport; specific organs
Erythrocytes:
  chloride fluxes, 181
  effect of SITS, 181
  exchangeable potassium, 24
  ghosts:
    preparation, 85-86
    sodium-potassium pump, 89-92
    structure, 87
  membrane potential, 85
Escherichia coli:
  active transport:
    of potassium, 141
    in respiration-deficient mutants, 141
  potassium accumulation, 33
    energy for, 33
  potassium-binding protein, 33
  protein synthesis:
    potassium dependence, 141-142
Ethacrynic acid:
  effect on cochlear ion transport, 162-163
  effect on kidney, 156
  effect on toad bladder, 157
N-Ethylmaleimide:
  inhibition K^+-activated phosphatase, 88
Evolution, 2-4
  of oceans, 2
Exchange diffusion:
  of cations, 32
Exchangeable potassium:
  in erythrocytes, 24
  in intestine, 23-24
  in lungs, 23-24
  measurement, 23-25
  in muscle, 24
  percentage total body potassium, 24

FCCP:
  action in mitochondria, 132-135
Feldspars, 1
  cation binding, 3
  crystalline structure, 3
Fertilizers:
  potassium in, 4
Formycin triphosphate:
  reaction with sodium-potassium pump, 97-99
Formyltetrafolate synthetase:
  potassium requirement, 145
Fructokinase:
  potassium requirement, 139
Fungi:
  potassium uptake, 10-11
  see also *Neurospora crassa;* Yeasts
Furosemide, 172

Gating mechanism:
  conformational change in membrane, 68
  effect of calcium, 80-82
  effect of pH, 80-82
  gating currents:
    effect of TEA, 69
    effect of tetracaine, 69
    in nerves, 68
    in skeletal muscle, 68

Gating mechanism (Cont'd)
  in nerves:
    in action potential, 59-61
      alpha position, 60
      beta position, 60
      rate of closing, 60
      rate of opening, 60
    in Purkinje fibers, 80-82
    in skeletal muscle:
      in action potential, 62-66
Glauconite, 2
Glucuronidase:
  accumulation during potassium deprivation, 144
Glutathionine:
  synthesis:
    potassium requirement, 146
Glycine transformylase:
  potassium requirement, 145
Gramicidin:
  action in mitochondria, 135
Griffithsia:
  active transport of potassium, 9
  intracellular potassium:
    activity coefficient, 37-38
    transmembrane electrochemical gradients, 9
Guinea pig *Taenia coli:*
  equilibrium potentials, 42
  ion permeability coefficients, 42-43

Halicystis ovalis:
  active transport of chloride, 10
  active transport of sodium, 10
Halicystis parvula:
  active trasnport of chloride, 10
Heart:
  failure:
    decrease in myocardium potassium, 20
  potassium concentration, 24
Helix aspersa:
  large neurons:
    potassium permeability, 55
Hepatocytes:
  membrane potential, 85
Hodgkin-Huxley model of conductance control:
  applied to cardiac muscle, 73
  applied to muscle, 65
  applied to nerves, 59-61
  gating mechanism, 59-61
Hydrolases, potassium requiring, 139
Hypercapnia, *see* Acidosis, respiratory
Hyperkalemia, 175

Igneous rocks:
  potassium content, 1-2
  weathering of, 1-2
Illite:
  formation of, 3
Inosinic acid ribotide:
  synthesis:
    potassium requirement, 145
Insulin:
  reduction of potassium fluxes, 104
  stimulation of (Na^++K^+)-ATPase, 105
  stimulation of sodium-potassium pump, 102, 104-106
Intestine:
  exchangeable potassium, 23-24
Iodoacetate:
  inhibition of sodium-potassium pump, 102
Iodonaphthyl-1-azide:
  labelling of (Na^++K^+)-ATPase, 88
Ion-selective channels:
  membrane theory of K^+ accumulation, 38-39
  see also Gating mechanism; Potassium channels; Sodium channels
Ionophore A23187:
  effect on potassium conductance, 54-55
Ischemia, local:
  decrease in intracellular potassium, 17, 20
Isoprenaline:
  stimulation of (Na^++K^+)-ATPase, 106
  stimulation of sodium-potassium pump, 105
Isoproterenol:
  stimulation of sodium-potassium pump, 102

Kaolinite, 1
Kidney:
  during dietary potassium depletion, 175

Kidney (Cont'd)
  distal tubules:
    $(Na^+ + K^+)$-ATPase, 168
    effect of amiloride, 171
    K^+-activated phosphatase, 168
    potassium concentration in fluid, 167
    potassium fluxes, 168-169
    sodium-potassium pump, 168
    effect of acidosis, 170
    effect of alkalosis, 170
    effect of diuretics, 156, 167, 171-172
    effect of potassium deprivation, 144-145, 170
    potassium concentration, 24
    potassium fluxes, 167-172
      effect of diuretics, 171-172
  proximal tubules:
    potassium fluxes, 167
  in respiratory acidosis, 175
Koeffoed-Johnsen Ussing model of transepithelial transport, 152-154
  experimental evidence, 158

Ligases, potassium requiring, 139
Lithium:
  blocking of potassium channels, 52
Liver:
  active transport, 112
  potassium concentration, 24
  protein synthesis:
    potassium dependence, 142
Lungs:
  exchangeable potassium, 23-24
Lysosomes:
  accumulation during potassium deprivation, 144

Macrotetralides, 148
Magnesium:
  role in protein synthesis, 141
  role in sodium-potassium pump, 91
Mammals:
  potassium requirement, 143-145
Membrane theory of K^+ accumulation, 38-57
  ion-selective channels, 38
  size distribution, 38
Mepyramine:
  effect on skeletal muscle, 103
Metamorphic rocks:
  potassium content, 2
Methionine:
  activation:
    potassium requirement, 146-147
Micas:
  biotite, 1
  cation binding, 3
  crystalline structure, 3
  ion exchange properties, 4-5
  potassium depletion, 3-4
  weathering of, 1-2
  white muscovite, 1
Mice:
  effect of potassium deprivation, 143-145
Michaelis-Menten kinetics:
  applied to active transport, 12-14
Microelectrodes, potassium selective:
  containing valinomycin, 28
  measurement of intracellular potassium, 25-28
  of potassium selective glass, 25-26
  with potassium selective liquid ion exchanger, 26-28
Mitochondria:
  cation concentrations, 121
  chemiosmotic hypothesis, 124-131
  contraction, 132
  effect on intracellular pH, 177
  generation of ATP, 124-131
  mammalian:
    effect of potassium deprivation, 144-145
  membrane potential, 121-124
    effect of metabolic state, 122-123
    effect of valinomycin, 123
  potassium fluxes, 120-135
    effect of gramicidin, 135
    effect of inorganic phosphate, 129
    effect of nigericin, 133-135
    effect of uncoupling agents, 132-135
    effect of valinomycin, 120-121
    energy requirements, 121
    role of inner membrane, 121
  proton pump, 120, 124-131
    effect of uncoupling agents, 132-135

Mitochondria (Cont'd)
  submitochondrial particles:
    cation fluxes, 135
    effect of nigericin, 135
    effect of uncoupling agents, 135
    effect of valinomycin, 135
  swelling, 131-132
    potassium movement, 132
  in transepithelial transport, 159
  uncoupling agents, 132-135
  volume control, 131-132
Mitotic apparatus:
  potassium dependence of synthesis, 142-143
Monactin, 148
Montmorillonite:
  formation of, 3
  ion exchange properties, 4-5
Muscle:
  exchangeable potassium, 24
  intracellular potassium:
    activity coefficient, 37
  intracellular sodium:
    activity coefficient, 37
    potassium concentration, 24
  see also Cardiac muscle; Resting muscle; Skeletal muscle; Smooth muscle; Vascular muscle
Muscovite:
  crystalline structure, 4
  white, 1
Myotonia congenita:
  in goats, 62
  in man, 62

Nerves:
  action potential, 58-61
    Hodgkin-Huxley model of conductance control, 59-61
    theoretical, 61
  electrogenic sodium pump, 102-103
  gating currents for potassium channels, 66-70
  intracellular pH, 181
    effect of cyanide, 181
    effect of DNP, 181
  intracellular potassium:
    activity coefficient, 37
  intracellular sodium:
    activity coefficient, 37
    maintainance of intracellular pH, 179
Neurospora crassa:
  intracellular cation concentrations, 10
  membrane potential, 10-11
  potassium uptake, 10-11
  in respiration-deficient mutant, 11
Nigericin:
  action in mitochondria, 133-135
  effect on submitochondrial particles, 135
Nitella translucens:
  active transport:
    of chloride, 6-7
    of potassium, 6-7
    of sodium, 6-7
  cytoplasmic potassium concentration, 6
  partial ion conductances, 8-9
  passive accumulation of potassium, 8
  total membrane conductance, 8-9
  transmembrane electrical gradients, 6
  transmembrane ionic gradients, 6
  vacuolar potassium concentration, 6
Nonactin, 148
Nystatin:
  effect on epithelial tissue, 154-155

Oceans:
  evolution of, 2
  potassium content, 2
Ouabain:
  binding in toad bladder, 153-154
  diuretic action, 171
  effect on *Cecropia* midgut tissue, 160
  effect on active transport in yeast, 113
  effect on intracellular pH, 181-182
  effect on toad bladder, 157
  inhibition of sodium-potassium pump, 89, 102
Oxaloacetate decarboxylase:
  potassium requirement, 139
Oxidoreductases, potassium requiring, 139

Pacemaker cells of heart:
  action potential, 73

pH:
  effect on gating, 80-82
pH, intracellular:
  effect of ammonia, 179
  effect of cell compartmentalization, 177
  effect of ouabain, 181-182
  effect of SITS, 181-182
  maintenance of:
    by active transport, 179, 181
    by sodium-potassium pump, 182
  measurement of, 176-177
    by distribution of weak acids and bases, 177
    by DMO method, 177
    with microelectrodes, 176-177
  in muscle, resting, 176-181
  in nerves, 179
  in red blood cells, 179
  role of bicarbonate, 176
  see also Acidosis; Alkalosis
Phosphatase, acid:
  accumulation during potassium deprivation, 144
Phosphatase, K^+-activated, 88, 147
  inhibition by N-ethylmaleimide, 88
  inhibition by P_1, 99
  in kidney, 168
  magnesium dependence, 99
  in red blood cells, 88
  relationship to sodium-potassium pump, 99-100
  relationship to sodium exchange, 100
Phosphate uptake:
  in bacteria, 114
  in yeast, 113-114
Phosphotransferases, potassium requiring, 139
Plants:
  potassium concentration in sap, 11
  potassium uptake, 3, 11-14
  see also Root hairs
Plasma, potassium concentration, 17, 24
Polyribosomes:
  potassium dependence, 140-143
  in protein synthesis, 140-143
Potash, 1
Potassium:
  accumulation:
    energy requirements, 32-33
    in freshwater algae, 8
    passive, 102-103
    transepithelial, 156
    see also Active transport; Association-induction hypothesis
  in acidosis, 175
  and action potential:
    of cardiac muscle, 70-80
    of nerves, 58-83
    of skeletal muscle, 62-66
    see also (Na^+ + K^+)-ATPase; Sodium-potassium pump
  activation of sulfhydryl compounds, 146-147
  active transport:
    in Cecropia midgut tissue, 160, 163-165
    in cochlea, 160-165
    in Griffithsia, 9
    in mitochondria, 120-135
    in Nitella translucens, 6-7
    in root hairs, 11-14
    in yeast, 13, 112-113
  in bacteria:
    regulation of osmotic pressure, 114
  conductance:
    effect of calcium, 54-55
    effect of ionophore A23187, 54-55
    effect of metabolic exhaustion, 54
    effect of valinomycin, 54
    factors increasing, 54-55
  deprivation in mammals, 143-145
    acid phosphatase accumulation, 144
    effect on kidney, 144, 170
    enzyme changes, 143-145
    glucuronidase accumulation, 144
    increase in lysosomal enzymes, 144
    in mitochondria, 144-145
  in environment:
    in clay-humus complexes, 2-4
    in igneous rocks, 1-5
    in metamorphic rocks, 2
    in oceanic sediments, 2
    in oceans, 2
    percent of lithosphere, 1
    reservoir, 4
    in sedimentary rocks, 2
    in soil water, 3
    in soils, 1-5
  enzymes requiring, 138-139, 145-148

Potassium (Cont'd)
  exchangeable, see Exchangeable potassium
  in fertilizers, 4
  fixation by marine organisms, 2
  fluxes:
    in acidosis, 182-185
    across kidney tubules, 167-169
    in alkalosis, 182-185
    in cochlea, 162-163
    in epithelial tissue, 159
    in mitochondria, 120-135
    in salivary glands, 165
  intracellular:
    concentration and membrane potential, 182
    measurement with ion selective electrodes, 25-28
  intracellular activity coefficient:
    in algae, 37-38
    in amphibian oocytes, 37
    in liver, 37
    in muscle, 37
  isotopes, 20-21, 23
  in maintenance of intracellular pH, 176-181
  permeability:
    of bacteria, 3
    of yeast, 3
  in repolarization of cardiac muscle, 70-80
  requirements for mammals, 143-145
  role in formation of mitotic apparatus, 142-143
  role in protein synthesis, 140-143
  role in purine biosynthesis, 145
  tissue concentrations:
    decrease in disease, 17, 20
    see also Specific organs
  total body, see Total body potassium
  uptake:
    by algae, 5-10
    in bacteria, 3, 114
    effect of calcium, 8
    energy requirement, 11
    in epithelial tissue, 154-155
    in fungi, 10-11
    in higher plants, 11-14
    in mitochondria, 120-135
    in *Neurospora crassa*, 7, 10-11
    in plants, 3, 11-14
    in rabbit bladder, 155
    in root hairs, 11-14
    in yeast, 11, 112-113
Potassium channels:
  blockage by caesium, 79-80
  dimensions, 52
  effect of aminopyridine, 53
  effect of barium, 52-54
  effect of caesium, 51-53
  effect of lithium, 52
  effect of metabolic exhaustion, 54
  effect of rubidium, 51
  effect of tetracaine, 69
  effect of tetraethylammonium ion (TEA), 52, 69
  gating currents, 66-70
Potassium pump, electrogenic:
  in *Cecropia* midgut tissue, 164-165
  in cochlea, 162
Potassium selective microelectrodes, see Microelectrodes, potassium selective
Potassium-binding protein:
  from *Escherichia coli*, 33
  from skeletal muscle, 33
Primaeval broth, 3
Propranolol:
  effect on cardiac muscle, 183
Protein synthesis:
  potassium dependence, 140-143
  in *Bacillus amyloliquifaciens*, 140
  in *Bacillus subtilis*, 140
  in *E. coli*, 141-142
  in liver, 141
  of polysome formation, 140-143
  in sea urchin eggs, 142
  synergism with magnesium, 141
  in yeast, 141
Proton fluxes:
  in resting muscle, 177
Proton pump:
  effect of ouabain, 181-182
  in mitochondria, 120, 124-131
    chemiosomotic hypothesis, 124-131
    effect of nigericin, 133-135
    effect of uncoupling agents, 132-135
  effect on intracellular pH, 177
  role in maintenance of intracellular pH, 176-181
  in yeast, 112-114

Purines:
　biosynthesis:
　　in *Aerobacter aerogenes*, 145
　　formyltetrafolate synthetase, 145
　　in plants, 145
　　potassium requirement, 145
Purkinje fibers:
　action potential, 75-80
　anomalous rectification, 75-76
　delayed rectification, 76
　effect of caesium, 79-80
　gating mechanism, 80-82
　pacemaker activity, 77-79
Pyruvate kinase:
　potassium dependence, 139

Rats:
　effect of potassium deprivation, 143-145
Red blood cells:
　$(Na^+ + K^+)$ -ATPase, 89-100
　high potassium, 93-96
　HK-LK dimorphism:
　　development during cell maturation, 95
　　genetic determinants, 94
　　relationship to M-L blood group antigen, 94-95
　intracellular pH, 179
　low potassium, 93-96
　potassium conductance:
　　effect of calcium, 54-55
　　effect of ionophore A23187, 54
　　effect of valinomycin, 54
Redox dyes:
　effect on active transport in yeast, 113
Renal medulla:
　$(Na^+ + K^+)$ -ATPase, 87-88
Resting muscle:
　intracellular pH, 176-181
　　effect of cell compartmentalization, 177
　　measurement of, 176-177
　　role of bicarbonate, 176
　　role of mitochondria, 177
Ribosomes:
　membrane-bound, 140
　effect of potassium, 140-143
Rocks:
　igneous, *see* Igneous rocks
　metamorphic, *see* Metamorphic rocks
　sedimentary, *see* Sedimentary rocks

Root hairs:
　active transport of potassium, 12-14
　intracellular potassium concentration, 11
　membrane potential, 11
　potassium uptake, 11-14
　proton excretion:
　　effect on soil potassium, 3-4
　　in weathering process, 2
Roots, excised:
　potassium uptake, 14
Rotenone:
　effect on kidney, 156
　effect on toad bladder, 157
Rubidium:
　blocking of potassium channels, 51-52

Saliva, 165
　secretion:
　　initiation of, 166
Salivary glands:
　active transport, 166
　$(Na^+ + K^+)$ -ATPase, 166
　effect of acetylcholine, 166
　effect of parasympathetic stimulation, 166
　potassium fluxes, 165
　　effect of carbachol, 165
　potassium secretion, 165-167
　　role of alpha-adrenergic receptors, 165
　　role of cholinergic receptors, 165
　sodium pump, electrogenic, 166
Sea urchin eggs:
　formation of mitotic apparatus, 142-143
　　potassium dependence, 142-143
　protein synthesis, 142-143
　　potassium dependence, 142-143
Sedimentary rocks:
　potassium content, 2
Silkworm, *see* Cecropia
SITS, *see* 4-Acetamido-4′-isothiocyano-stilbene-2,2′-disulfonic acid
Skeletal muscle:
　action potential, 62-66
　　compared to cardiac muscle, 70
　active transport, 100-102
　　effect of inervation, 102-103
　afterpotential, 63
　anomalous rectification, 41, 45-51
　application Hodgkin-Huxley model, 65

198 Index

Skeletal muscle (Cont'd)
biopsy for assessment of body K^+ status, 18-20
delayed rectification, 62-63
effect of adrenaline, 105
effect of insulin, 105
effect of isoprenaline, 105-106
effect of local anesthetics, 103
effect of mepyramine, 103
electrogenic sodium pump, 102-103
equilibrium potentials:
   role of internal membrane system, 66
gating currents for potassium channels, 66-70
intrafiber ion diffusion, 34-37
   effect of compartmentalization, 34-35
ion permeabilities, 39-41
ion permeability coefficients, 44
membrane potential, 40-41
metabolic acidosis, 182-185
potassium accumulation, 33-37
   effect of pH, 33
potassium concentration, 18-20
   pale phasic fibers, 19
   red tonic fibers, 19
potassium conductance:
   factors increasing, 54-55
potassium-binding protein, 33
relationship between sodium and potassium, 19-20
respiratory acidosis, 182-185
sodium-potassium pump, 100-102
T-system, 65
   in anomalous rectification, 47-50
Skin:
   potassium concentration, 24
Skin, frog:
   active transport, 151-152
   ion permeabilities, 152
   voltage steps across, 152
Smooth muscle:
   equilibrium potentials, 42
   guinea-pig *Taenia coli*, 42-43
   ion permeability coefficients, 42-44
   sodium-potassium pump, 107-109
Sodium:
   and action potential:
      in cardiac muscle, 70-80
      of nerves, 58-83

   of skeletal muscle, 62-66
active transport:
   in frog skin, 151
   in *Halicystis ovalis*, 10
   in kidney, 168-171
   in *Nitella translucens*, 6-7
   in salivary glands, 166
   in toad bladder, 153-154
fluxes:
   in cochlea, 162-163
intracellular activity coefficient:
   in algae, 37-38
   in amphibian oocytes, 37
   in liver, 37
   in muscle, 37
   in maintenance of intracellular pH, 176-181
uptake:
   in epithelial tissue, 154-159
Sodium channels:
   blockage with tetrodotoxin, 63
   effect of tetrodotoxin, 73
Sodium pump, electrogenic, 102-103
   in cochlea, 162
   effect of local anesthetics, 103
   effect of mepyramine, 103
   in epithelial tissue, 155
   in nerves, 102-103
   in salivary glands, 166
   in skeletal muscle, 102-103
   in sweat glands, 167
   see also ($Na^+ + K^+$)-ATPase; Sodium-potassium pump
Sodium-potassium pump:
   ATP requirement, 89-92
   ATP-affinities, 97
   cardiac muscle, 106-107
      effect of cardiac glycosides, 106-107
   in cochlea, 162
      location, 163
   conformational changes, 96-99
   dephosphorylation of membrane proteins, 92-93
   in dialyzed squid axon, 109-112
   effect of 2,3-dimercaptopropanol, 91
   effect of arsenite, 91
   effect of ouabain, 89
   K^+-K^+ exchange, 92-93
   ouabain inhibition, 93

Sodium-potassium pump (Cont'd)
  Na^+-Na^+ exchange, 89-92
  in kidney, 168
  kinetics of ATP hydrolysis, 97-99
  phosphorylation of membrane proteins, 1-92
  reaction with formycin triphosphate, 97-99
  in red blood cells, 89-92
  relationship to K^+-activated phosphatase, 99-100
  reversibility, 89
  role in maintenance of intracellular pH, 176-182
  role of magnesium, 91
  in skeletal muscle, 100-102
    energy requirement, 100-101
    inhibition by iodoacetate, 102
    inhibition by ouabain, 102
    stimulation by insulin, 102
    stimulation by isoproterenol, 102
  in smooth muscle, 107-109
  stimulation by beta-adrenergic agents, 104-106
  stimulation by catecholamines, 105-106
  stimulation by insulin, 104-106
  in vascular muscle, 108-109
Soils:
  potassium content, 1
  potassium exchange, 3-4
Soil water:
  potassium concentration, 3
Squid axon, dialyzed:
  effect of strophanthidin, 110
  potassium transport, 109-112
  preparation of, 109
  sodium transport, 109-112
Strophanthidin:
  effect on cardiac tissue, 107
  effect on dialyzed squid axon, 110
  inhibition of $(Na^+ + K^+)$-ATPase, 89
Submitochondrial particles:
  cation fluxes, 135
  effect of nigericin, 135
  effect of uncoupling agents, 135
  effect of valinomycin, 135
Succinic dehydrogenase:
  potassium requirement, 144
Sulfhydryl compounds:
  activation:
    potassium requirement, 146-147
Sweat glands:
  potassium transport, 167
  sodium pump, electrogenic, 167

Tetracaine:
  effect on potassium channels, 69
Tetraethylammonium ion:
  blocking of potassium channels, 52
Tetrodotoxin:
  effect on sodium channels, 63
  in cardiac muscle, 73
Total body potassium:
  analysis of tissue biopsies, 18-20
  of cadavers, 22
  exchangeable potassium measurement, 23-25
  extracellular fraction, 18
  on fat-free tissue basis, 18
  human female, 18
  human male, 18
  intracellular fraction, 18
  whole body counters, 20-23
    controls for human subjects, 21
    equipment, 21
    isotopes used, 20-21
Transepithelial transport:
  in *Cecropia* midgut tissue, 160, 163-164
  in cochlea, 160-165
  in kidney, 167-172
  Koeffoed-Johnsen Ussing model, 152-154
  in salivary glands, 165-167
  serosal membrane:
    active transport of potassium, 156
  in sweat glands, 167
  in toad bladder, 152-153
Transformylation:
  potassium requirement, 145
Translation, *see* Protein synthesis
Tryptophanase, potassium requirement, 139

Uncoupling agents:
  effect on mitochondria, 132-135
  effect on submitochondrial particles, 135
Urine:
  formation:
    effect of potassium deprivation, 144
  potassium concentration, 167

Valinomycin:
  effect on membrane potential of mitochondria, 123-124
  effect on mitochondria, 120-135
  effect on potassium conductance, 54
  effect on submitochondrial particles, 135
  in potassium selective microelectrodes, 28
Vascular smooth muscle:
  sodium-potassium pump, 108-109
Vasopressin:
  effect on epithelial tissue, 158
Ventricle atrium of heart:
  action potential, 73
Vermiculite:
  exchangeable potassium, 4
  formation of, 3

Weathering:
  chemical, 1-2
  mechanical, 1
  release of potassium by, 1-2
  role of plants, 2
Whole body counters, *see* Total body potassium; Whole body counters

Yeasts:
  active transport, 11, 112-114
    effect of redox dyes, 113
    of potassium, 13, 141
    in respiration-deficient mutants, 141
  effect of potassium deficiency, 139
  intracellular potassium concentration, 3
  phosphate uptake, 113-114
  potassium uptake, 3, 11, 112-114
    effect of ouabain, 113
    effect of pH, 112
    inhibitors, 113
    in respiration-deficient mutant, 113
  protein synthesis:
    potassium dependence, 141
  proton pump, 112-114